WORLD BANK STUDY

Power and People: The Benefits of Renewable Energy in Nepal

Sudeshna Ghosh Banerjee
Avjeet Singh
Hussain Samad

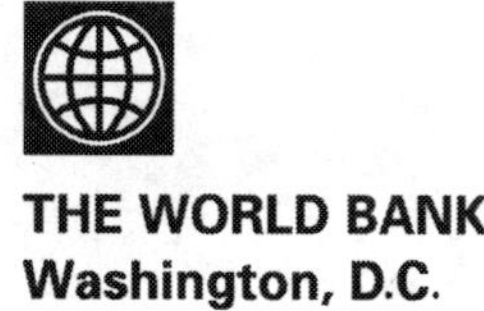

THE WORLD BANK
Washington, D.C.

South Asia Energy Unit/Sustainable Development Department/The World Bank
1818 H Street, NW
Washington, DC 20433
Telephone: 202-473-1000
Internet: www.worldbank.org

1 2 3 4 14 13 12 11

ISBN: 978-0-8213-8779-5
eISBN: 978-0-8213-8789-4
DOI: 10.1596/978-0-8213-8779-5

Library of Congress Cataloging-in-Publication Data

Power and people: the benefits of renewable energy in Nepal / Sudeshna Ghosh Banerjee, Avjeet Singh, Hussain Samad.
p. cm.
"June 2010."
Includes bibliographical references.
ISBN 978-0-8213-8779-5 -- ISBN 978-0-8213-8789-4 (electronic)
1. Rural electrification--Nepal. 2. Renewable energy--Nepal. 3. Rural development--Nepal. I. Banerjee, Sudeshna Ghosh, 1973- II. Singh, Avjeet. III. Samad, Hussain A., 1963-
HD9688.N352.P69 2011
333.79'4095496--dc22 2011014933

Contents

Boxes

Figures

Tables

Acronyms and Abbreviations

AEPC	Alternative Energy Promotion Center
AP	Actual Pay
BSP	Biogas Support Program
CAGR	Compound Annual Growth Rate
CDM	Clean Development Mechanism
CM	Community Mobilizer
CMC	Community Mobilizer Coordinator
CO	Community Organizations
CREF	Central Rural Energy Fund
CRT/N	Center for Rural Technology, Nepal
DDC	District Development Committee
DDG	Decentralized Distributed Generation
DEES	District Energy & Environment Section
DGIS	Netherlands Directorate-General of Development Cooperation
DKK	Danish Krone
EC	Energy Conservation
EDO	Energy Development Officer
ESAP	Structural Adjustment Program
ESMAP	Energy Sector Management Assistance Program
ESW	Economic and Sector Work
FF	Field Facilitator
FG	Functional Groups
FIT	Feed-in Tariff
GBI	Generation-based Incentives
GEF	Green Energy Fund
GDP	Gross Domestic Product
GHG	Greenhouse Gas
GI	Gastrointestinal
GKD	The Kingdom of Denmark
GOA	Ghatta Owners Association
GoN	Government of Nepal
GSIA	Gender and Social Inclusion Advisor
HH	Household
HRDA	Human Resources Development Advisor
ICS	Improved Cooking Stoves
IFI	International Financial Institutions
IGA	Income Generating Activity
INPS	Integrated Nepal Power System
IPP	Independent Power Producer

ISPS	Institutional Solar PV System
IWMP	Improved Water Mill Program
IT	Information Technology
KfW	Kreditanstalt für Wiederaufbau
KPI	Key Performance Indicator
LPA	Livelihoods Promotion Adviser
LPG	Liquid Petroleum Gas
LPO	Local Partner Organisation
M&E	Monitoring & Evaluation
MCO	Monitoring & Communication Office
MDG	Millennium Development Goals
MH	Micro Hydro
MIRMS	Management Information Reporting and Monitoring System
MIS	Management Information System
MISA	Management Information Systems Associate
MoE	Ministry of Energy
MSW	Municipal Solid Waste
NEA	Nepal Electricity Authority
NGO	Non-government Organizations
NPM	National Program Manager
NWP	National Water Plan
O&M	Operation and Maintenance
PCC	Plant Completion Certificate
PCR	Plant Completion Report
PLF	Plant Load Factor
PPS	Probability Proportionate to Size
PSM	Propensity Score Matching
PV	Photovoltaics
PVPS	Solar Photovoltaic Pumping System
PwC	PricewaterhouseCoopers
REDA	Rural Energy Development Advisor
REDP	Rural Energy Development Program
REF	Rural Energy Fund
RELD	Rural Energy Livelihoods Development
REO	Renewable Energy Obligation
REP	Renewable Energy Project
RESD	Regional Energy Systems Development
RESS	Renewable Energy Sector Support
RET	Renewable Energy Technology
RPS	Renewable Portfolio Standard
RRESC	Regional Renewable Energy Service Center
SHS	Solar Home Systems
SO	Support Organization
SODP	Strategic and Organizational Development Plan
SRESDA	Senior Rural Energy System Development Advisor
SNV/N	Netherlands Development Organisation/Nepal

SWOT	Strength, Weakness, Opportunity, Threat
SWH	Solar Water Heater
UNDP	United Nations Development Program
VC	Vulnerable Community
VDC	Village Development Committee
WB	World Bank
WTP	Willingness to Pay

Acknowledgments

This report was drafted by a team led by Sudeshna Ghosh Banerjee and comprising of Avjeet Singh and Hussain Samad, based on a series of reports produced by the Consulting Team of PricewaterhouseCoopers (PwC), India. The PwC team included Chandrashekar Iyer, Rakesh Jha, Manoj Mania, and S. Johnny Edward. The work was carried out under the overall guidance of Salman Zaheer (Sector Manager, SASDE) and Michael Haney (Senior Energy Specialist, SASDE). The team thanks peer reviewers Shahidur Khandker, Cindy Suh, and Luis Alberto Andres, who provided many thoughtful comments and constructive observations. Funding from ESMAP and support from Amarquaye Armar and Rohit Khanna (Program Manager, ESMAP) is gratefully acknowledged.

Finally and most important, the team owes a debt of gratitude to Alternative Energy Promotion Center of Nepal, particularly Narayan Chaulagain (Executive Director), Govind Pokharel (Former Executive Director), Kiran Man Singh (National Program Manager, REDP), and Bharat Poudel (M&E Officer) who coordinated the nodal working group as well as communication with World Bank staff and consultants. Special thanks to Marjorie K. Araya (ESMAP) for carrying out comprehensive copyediting and coordinating production jointly with EXTOP colleagues, Anna Socrates and Aziz Gokdemir.

Executive Summary

A large section of the Nepalese population is deprived of electricity coverage despite huge hydropower potential, particularly in the rural areas. About 63 percent of Nepalese households lack access to electricity and depend on oil-based or renewable energy alternatives. The disparity in access is stark, with almost 90 percent of the urban population connected, but less than 30 percent of the rural population. Nepal has about 83,000 MW of economically exploitable resources, but only 650 MW have been developed so far.

Decentralized service delivery in the form of renewables such as micro-hydro and solar can fill more of the gap in rural areas. Traditional water wheels (*gharats*) provide electricity to 7 percent of the rural population, equivalent to about 800,000 people. The 10th Five-Year Plan sets a target of generating electricity equivalent to 10 MW from micro-hydro schemes and access to off-grid electrification for 12 percent of the population.

The primary institutional responsibility of providing energy access in rural areas using renewable technologies falls on Alternative Energy Promotion Center (AEPC). AEPC aims to energize rural households and kick-start economic activities in the project areas. Since its inception in 1996, it has emerged as one of the world's leading proponents of community- and private-sector–led expansion of renewable energy technologies in rural areas. It is a government institution falling under the oversight of the Ministry of Environment and operating with semi-autonomous status. Its mandate is policy and plan formulation, resource mobilization, coordination, and quality assurance. Its programs primarily relate to micro/mini-hydro and improved water mills, solar energy (PV and thermal), biogas, improved cooking stoves, wind energy, and geothermal (figure 1).

Figure 1: Number of households benefiting from renewable technologies

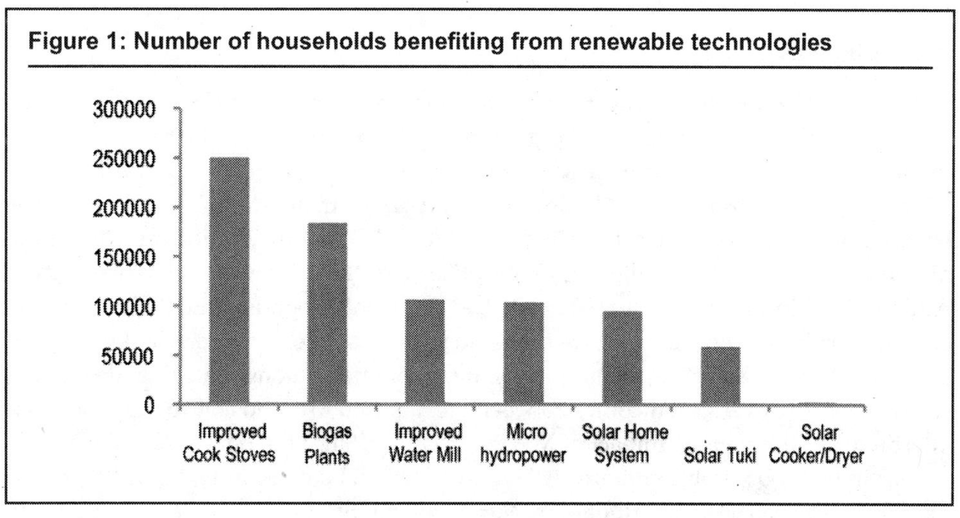

Source: AEPC.

This study has been designed to organize an evaluation system that measures the impact of micro-hydro installations on rural livelihoods and to establish a monitoring system for AEPC to continually measure the results of the renewable energy programs against the targets. Given AEPC's high dependence on donor and Government funds, the need to develop a system that provides information on a wide range of technical, operational, and financial parameters is similarly high. The integrated M&E system encompasses all of AEPC's programs in micro-hydro (MH), solar, biomass, improved water mills, and biogas, and builds its capacity to execute it.

A set of key performance indicators (KPIs) has been selected across the results chain of monitoring framework to facilitate informed decision-making. The indicators are monitored throughout implementation to assess progress toward program objectives. The indicators have been classified as follows:

- *Inputs:* include financial, human, and other technical resources mobilized to support activities undertaken by AEPC. Examples include funds utilized, value of land, and number of awareness-generation programs for community mobilization.
- *Outputs:* include the interventions by AEPC in such technical aspects as systems installed and numbers of households connected; capacity building such as number of staff trained; and community participation such as number of community organizations.
- *Outcomes:* measure the uptake, adoption, and use of services by project beneficiaries. They are assessed on various parameters including access, affordability, operating efficiency, reliability, and financial sustainability.
- *Impacts:* include long-term outcomes near or at the end of the results chain, such as the effect of provision of electricity on employment, health, education, and income. Since this information needs to be collected at a household level, it is suggested to be collected every 2–3 years based on a survey.

Impact indicators

The monitoring framework incorporates the final impacts of electrification on households and businesses, evaluated using a primary household and enterprise survey. In the survey, 2,500 households were selected, of which 1,500 are treated (MH connected), 571 are in control groups from MH locations and 429 from non-MH (control) locations. For the enterprise survey, 70 enterprises were selected, 34 in MH locations and 36 in non-MH locations. These households and enterprises have been provided electricity under AEPC's flagship program—Rural Energy Development Program (REDP). The welfare outcomes of the MH households are compared with those of control households through a propensity score matching technique, the most common among matching techniques. The difference of average outcomes between the matched MH and control households is interpreted as the average impact of MH connectivity. Since the households are exactly similar in their observable characteristics except for MH connectivity, the difference in their outcomes has been contributed by MH connectivity.

The primary use of MH is lighting for the households, with about 10 percent for appliances. Biomass energy, fuel wood in particular, is by far the most dominant form of energy for rural households regardless of their MH connectivity. MH replaces or reduces the use of kerosene for lighting. The average electricity consumption of MH users is only about 1.9 kgOE/month, roughly 4 hours of daily use of three 60-watt bulbs. The consumption of MH electricity in rural Nepal is limited by a capacity ceiling that is good for just basic lighting and a limited number of low-wattage appliances or gadgets, so simultaneous use of all of them is often not possible.

Rural households spend about 11 percent of their income on energy, and the share of energy spending in total income goes down steadily as household income goes up. The poorest households spend 22 percent of their income on energy, the wealthiest households, 6.4 percent. Among MH users, high-income households consume more electricity than their lower income counterparts. But the trend in the average price paid (Rs/kWh) is not so consistent, although overall it goes down as income goes up. Therefore, high-income households not only consume more electricity than low-income households, they use electricity more optimally so as to pay a lower unit price. The richest households consume almost twice as much electricity as the poorest households but pay only 32 percent more.

Table 1: Expenditure of rural and MH households in Nepal

	Quintile 1	Quintile 2	Quintile 3	Quintile 4	Quintile 5	All
All households						
Energy expenditure (Rs./month)	738	829	861	973	1,055	875
Energy expenditure as a percentage of income	22	14.8	12.4	10.5	6.4	11.4
MH households						
Electricity expenditure (Rs./month)	54	58	58	70	71	62
Price paid (Rs./kWh)	3.32	3.05	3.18	2.97	2.19	2.83

Source: World Bank-AEPC survey 2009.

The lower bound of benefits to households measured in the form of consumer surplus is high. The consumer surplus estimates the benefits of households switching to MH, captured in the form of savings in the consumption of lighting energy. Households enjoy a surplus by switching to MH because the amount paid for it or actual pay (AP) is less than what they would have for kerosene lighting, which is a proxy for willingness to pay (WTP). AP is lower than WTP because the unit price for lighting intensity (in Rs./klumen-hr) is much cheaper for MH: kerosene is 400 times more expensive than MH when measured by lighting intensity. Cheaper prices enable MH households to enjoy more lighting than non-MH households—a consumer surplus of almost Rs. 700 per month.

Access to MH imparts a wide range of benefits to its users—improving household economic, education, health, and women's empowerment outcomes. MH access increases households' non-farm income by 11 percent and consumption expenditure by about 6 percent. Girls' completed schooling years go up by 0.24 grade because of MH connectivity. Women and children from the MH households suffer less from respiratory problems than their counterparts from non-MH households. Women's contraceptive prevalence, involvement in income-generating activities, and decision-making independence all go up because of the MH connectivity.

Table 2: Benefits of rural electrification to MH connected households

Outcome variables	MH-Connected HHs	Non-MH HHs	Difference	Propensity Score Matched Difference
Economic				
Non-farm income (Rs./capita/month)	865.2	629.2	0.353 (2.53)**	0.112 (1.91)*
Expenditure (Rs./capita/month)	1,456.2	1,263.1	0.039 (1.92)*	0.090 (3.26)**
Education				
Schooling-years completed Girls' completed schooling years	4.28	3.73	0.551 (1.97)*	0.240 (1.65)*
Evening study (minutes/day)				
Boys' time spent in evening study (*minutes/day*)	50.1	33.9	16.2 (3.63)**	7.7 (2.32)**
Girls' time spent in evening study (*minutes/day*)	39.7	30.0	9.7 (2.19)**	12.0 (5.06)**
Health				
Adult women's respiratory problems	5.1	9.7	-4.62 (-0.91)	-3.4 (-3.22)*
Boys' respiratory problems	1.4	5.1	-3.63 (-1.75)*	-1.6 (-2.28)**
Girls' respiratory problems	1.3	8.2	-6.90 (-1.93)*	-6.1 (-2.82)**
Girls' *gastrointestinal (GI) problems*	0.3	1.7	-1.40 (-0.87)	-1.43 (-1.71)*
Women's Fertility and Empowerment				
Contraceptive prevalence rate	0.744	0.718	0.026 (0.59)	0.038 (2.78)**
Time-use in Income-generating activities (hours/day)	5.81	5.54	0.27 (1.99)**	0.19 (1.97) *
Study time (hours/day)	0.96	0.79	0.17 (1.13)	0.20 (1.86) *
Time in leisure activities (hours/day)	0.71	0.48	0.23 (4.25)**	0.21 (5.71)**
Independence in mobility type 1	0.569	0.354	0.215 (4.12)**	-0.013 (-0.36)
Independence in decision-making in fertility issues	0.844	0.726	0.117 (2.44)**	0.042 (1.85)*
Independence in decision-making in children's issues	0.942	0.921	0.021 (0.70)	0.027 (2.40)**

Note: * and ** refer to significance levels of 10 percent and 5 percent respectively. Figures in parentheses are t-statistics Type 1 mobility refers to frequent or occasional visits to places of friends, neighbors, and relatives. The table presents only results that are statistically significant.
Source: World Bank-AEPC survey 2009.

About 10 million kgs of CO_2 are saved every year by MH households in Nepal as they replace kerosene by MH electricity. Assuming that burning of one liter of kerosene emits 2.8 kg of CO_2, the difference in emissions between MH and non-MH households is significant—MH households emit about 3.6 kg less CO_2 per month than non-MH households. That amounts to a reduction of more than 10 million kg of CO_2 every year for MH households in Nepal, significant for a small country like Nepal where only 7 percent of the rural households have MH connectivity.[2]

The benefits from rural electrification are immense, but the complete potential will be realized only if service delivery is better. The average duration of outage is 9 hours per day, and it does not vary by the level of consumption. Besides power outages, voltage fluctuation affects almost 62 percent of MH users. Average monetary loss for voltage fluctuation is about 16 Rs. per month, and no clear pattern of losses can be established as electricity consumption goes up. MH households use a variety of alternative arrangements to cope with outages, among which kerosene lamps rank by far the highest. It clearly explains why MH households spend a fair amount of money on kerosene (more than 92 Rs. per month). For rural MH enterprises, the incidence of power outage is 100 percent. The average duration of outage is 8 hours, and it does not vary much across enterprise quintiles. More than half the enterprises report that they face voltage fluctuations. A majority use kerosene lamps, followed by emergency light. A number of MH enterprises use candles, and some stay in the dark. Measures taken by the enterprises are meant to satisfy their lighting needs only—no measures are taken to compensate for the loss of processing energy due to outages.

Input, output, and outcome indicators

The baseline data were gathered to establish a baseline for regularly tracking future improvements. The indicators are monitored throughout implementation to assess progress toward program objectives. The monitoring can be institutionalized in the form of a pyramid—with high-level strategic indicators focusing on outcomes and outputs at the top, facilitating the decision-making of the program managers (figure 2).

Figure 2: Development of key performance indicators

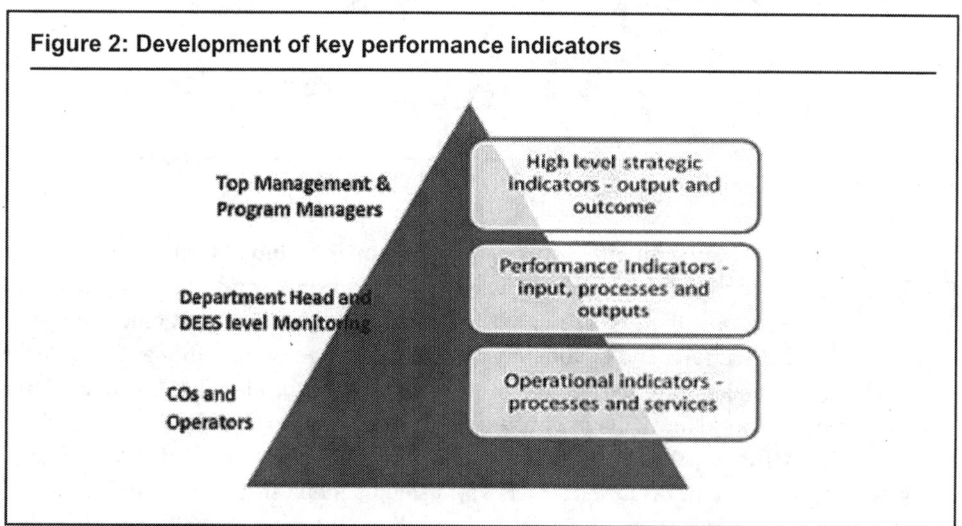

Source: Authors' elaboration.

The primary objective of the KPIs is to facilitate a framework for the government and the donors to assess the performance of AEPC on key parameters. But the true benefit of the indicators can be realized only when they are capable of being used as an organization-wide tool to monitor and track performance of AEPC's various units. Therefore, these KPIs have been operationalized by developing a robust MIS system. This enables the middle and top management of AEPC to have timely access to reliable and accurate information for informed decision-making. It helps AEPC monitor the development and implementation of its programs against set objectives.

The design of the MIS and the frequency of information flow has been based on the information needs of various levels of management within AEPC. The focus is not only on measuring day-to-day operational performance but also using information as a strategic planning tool. The periodicity of review varies depending upon the level of management targeted. In the redesigned system, information is captured at source with an operational review at the middle management level and a strategic review at the top management level.

- Monthly monitoring of operational data is by the field staff responsible for the respective functions.
- Trimester monitoring of performance parameters by the district heads or monitoring officers of respective programs.
- Annual, trimesterly, or monthly review of the program and strategic parameters by the top management for informed decision-making and reporting to the external stakeholders (figure 3).

Figure 3: Frequency of information flow

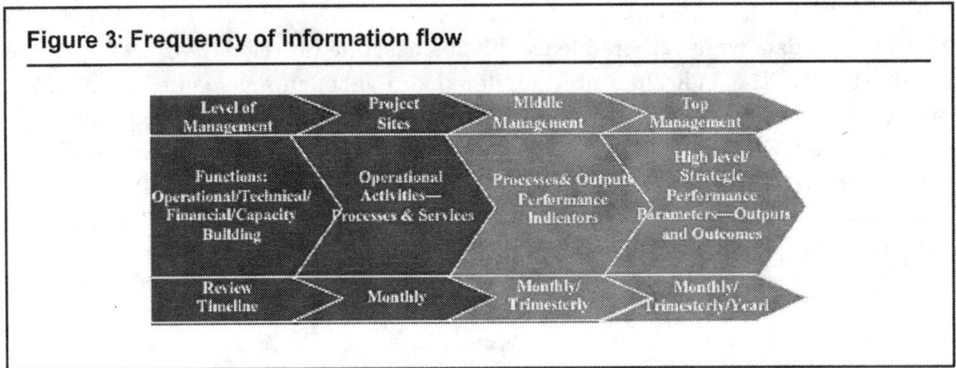

Source: Authors' elaboration.

A web-based Management Information Reporting and Monitoring System (MIRMS) will serve as the primary interface for all monitoring and analysis purposes. The system will support the decision-making capability of AEPC management by generating a set of MIS reports and evaluation dashboards based on monthly and quarterly program data for stakeholders. The system is designed to compile the performance parameters at project level and village development committee level and above for each of the programs. MIRMS will use the existing computer infrastructure of the AEPC server for hosting the system on the Internet. The users of the system can access it locally through a LAN or from any remote location through the Internet. With the central AEPC server at the head office of AEPC, consolidated data from all AEPC's programs would be populated periodically. Data would be either entered directly or extracted and uploaded in the system from different programs. MIRMS will interface with other external systems either through on-line or off-line mode to extract information captured through other databases. Structured and unified formats have been designed to facilitate ease in data gathering across various locations (figure 4).

Figure 4: Data flow structure

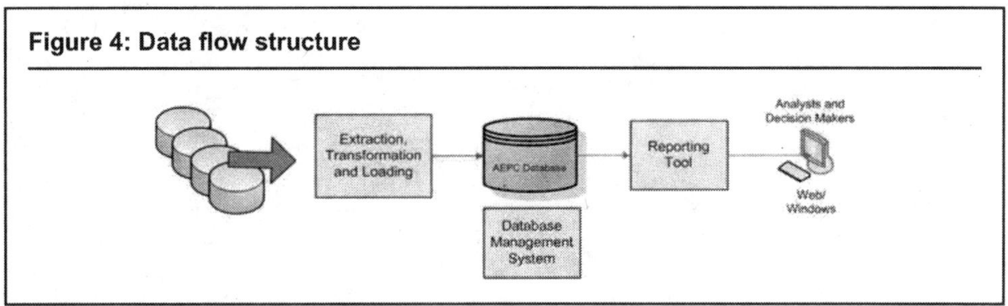

Source: Authors' elaboration.

Increasing coverage and using micro-hydro to the maximum

On the supply side, an MH plant is not viable without major subsidies. The cost of delivering MH to a household is about 6.86 Rs./kWh, about half from different forms of subsidy. The rest must be met by the revenue from the customers and savings mobilized by community (if any). MH customers pay a flat tariff based on a peak power purchased, as opposed to the actual consumption in kWh per month. Although the tariff for MH users varies to some extent, the average of the charged tariff is 1.5 Rs./month/watt. This may not be enough for many MH plants to break even.

On the demand side, lighting-only MH users (the majority of MH households) use electricity for only 4–5 hours during the evening, which results in a low load factor (about 33 percent), and thus indicates an inefficient use of electricity. Using MH for more diverse, especially income-generating, purposes may improve the load factor and make it more attractive. It will also lower the average effective price for consumption of MH electricity (now 2.8 Rs./kWh). In addition, the connection cost may be prohibitive—on average about 8,000 Rs. And unreliable service in the form of frequent and prolonged power outage and voltage fluctuations works as a deterrent to MH access for many households.

AEPC has to broaden its focus from installing the renewable facilities to ensuring that service is reliable and consumers are actually using it to the maximum load. Otherwise, these facilities will continue to be cost-ineffective and waste scarce resources. The M&E framework developed as part of this study can be an important tool to measure the progress of investment programs and to address the gaps in service delivery and devise appropriate solutions.

CHAPTER 1

A Long Road to Expanding Rural Access

Nepal is in an energy crisis. Nepal has about 703 MW of electricity generation capacity, comprising 53.41 MW thermal projects, 100 kW of solar projects and the rest from hydro power projects.[1] The peak demand in the country has gone up to 812.50 MW, and it is expected that the condition may further worsen in the years to come. Despite the difficult political environment for reforms and development activities, Nepal's GDP grew by 3.8 percent annually on average during FY2005–FY2009.[2] The energy requirement for the overall economy has increased many folds and will increase further in view of the expected future growth.

Access to electricity is low and inequitable, particularly in the poor rural areas. Nepal is one of the poorest countries in the world. About three-fourths of its households live in absolute poverty and depend largely on the natural environment to exist. About 70 percent of its 27 million people reside in rural areas and are engaged in subsistence farming. About 63 percent of households do not have access to electricity and depend on oil-based or renewable energy alternatives. Disparity in access is stark, with almost 90 percent of the urban population connected, but less than 30 percent of the rural. Rural households can pay up to 14 percent of their total household expenditures on electricity. Among the poor, the burden of electricity expenditures constitutes about 2 percent on average and about 6 percent maximum. Among the small population connected to electricity in rural areas, 86 percent pay for electricity.

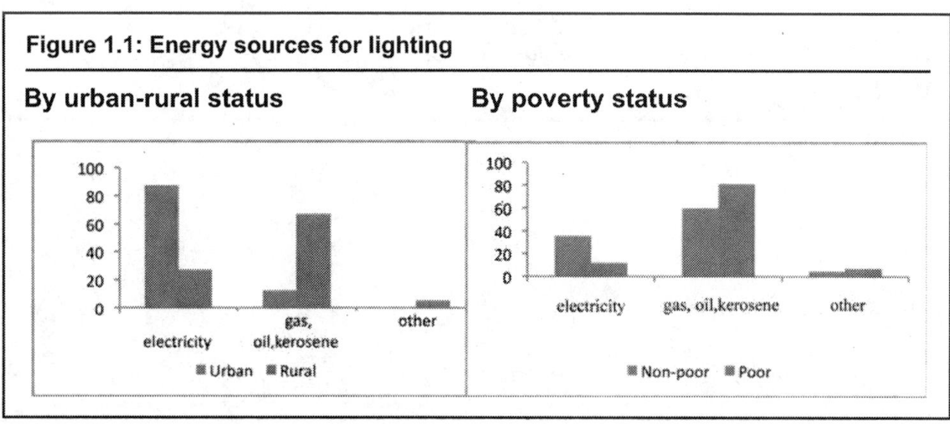

Figure 1.1: Energy sources for lighting

Source: Nepal Living Standards Survey, 2003/4.

There is significant unexploited hydro potential. Nepal has vast hydro resources, which present a source of potential wealth. Out of the theoretical potential of 83,000 MW, 42,000 MW is said to be economically viable in the present context. Despite this generation potential, only about 650 MW have been developed so far.

The geographical profile of Nepal does not allow grid extensions to remote rural areas due to high infrastructure cost and losses, lack of load, and poor returns compounded by constrained Government budgets. Development of off-grid renewable energy has been an important part of enhancing rural access. The rural population has to rely heavily on conventional sources of energy (biomass) and small micro-hydro schemes to meet power needs.

Decentralized electricity service delivery in the form of micro-hydro and solar photovoltaics (PV) are widely used renewable technologies for rural electrification in Nepal. The country is exposed to pico- and micro-hydro technologies. There is a sound presence of traditional water wheels (*gharats*) in Nepal. Currently, this source provides electricity to only 6–7 percent of the rural population. The 10th Five-Year Plan sets a target of generating electricity equivalent to 10 MW from micro-hydro schemes and access to off-grid electrification for 12 percent of the population. The GoN, in its 10th Five-Year plan (2003–2007), had set the ambitious target of electrifying up to 2,600 VDCs through the extension of the integrated national grid system. The energy needs of 1,000 additional VDCs were proposed to be met by decentralizing energy production systems as called for in the 10th Five-Year Plan. Due to the rugged mountainous terrain and scattered nature of human settlements, the national grid extension to these areas is difficult and not economical.

Table 1.1: Household beneficiaries of renewable energy

SN	Renewable Energy Type	Unit	Progress	Number of HHs benefited
Access to Electricity			6 % Population	
1	Biogas Plants	Number	184,000	184,000
2	Micro hydropower	kW	10,347	103,470
3	Solar Home System	Number	95,000	95,000
4	Solar Tuki	Number	59,120	59,120
5	Solar Cooker/Dryer	Number	2,500	2,500
6	Improved Cook Stoves	Number	250,000	250,000
7	Improved Water Mill	Number	3,552	106,560
	Total			800,650

Source: AEPC, 2008.

What is the policy framework and institutional structure for rural energy?

The electricity sector in Nepal is guided by various legislations and policies framed by the Government of Nepal (GoN). The Hydropower Development Policy of 2001, especially, places emphasis on rural electrification and the involvement of the private sector in the hydropower sector by offering a one-window policy and incentive packages.

- *Hydropower policy 2001*—"The policy aims to render support to the development of rural economy by extending the rural electrification and to implement small, medium, large and storage projects for hydropower development."
- *National water plan (NWP) 2005*—The plan encompasses programs in all strategically-identified output activities so that tangible benefits can be delivered to all the people in line with the basic needs. The plan also sets target for development of hydropower and other alternate energy resources in terms of capacity and reach to the people.
- *Rural energy policy (REP) 2006*—"The overall goal of this policy is to contribute to rural poverty reduction and environmental conservation by ensuring access to clean, reliable and appropriate energy in the rural areas."
- *Subsidy delivery mechanism 2006*—aims to ensure disbursement of subsidy in a cost effective and easy access manner. The mechanism places emphasis on effective expansion of the renewable (rural) energy market. For promotion of rural electrification, the mechanism specifies the eligible technologies like solar cooker, Solar PV systems, and Biomass and the criteria for disbursement of subsidy. AEPC will monitor and evaluate the performance of the program, including the implementing agency.
- *Subsidy for renewable energy 2006*—was formulated to make the already existing subsidy policy equitable, inclusive, and effective.
- *Community electricity distribution by law*—promotes the renewable energy sector by targeting these policies on important elements such as subsidy, encouragement to the private sector, and community-based developments, which are vital for the electrification of rural areas using renewable sources in general and hydro in particular.

A multitude of institutions govern the electricity sector in Nepal. The line ministry in charge of micro-hydro development is the Ministry of Environment (MoE). In its mandate for Hydropower Development, MoE is also charged with Rural Electrification. The Nepal Electricity Authority (NEA) under Ministry of Energy has traditionally been the most important implementing agency of renewable energy. NEA was created on August 16, 1985, under the Nepal Electricity Authority Act 1984 through the merger of the Department of Electricity in the Ministry of Water Resources, the Nepal Electricity Corporation, and related Development Boards. The primary objective of NEA is to generate, transmit, and distribute adequate, reliable, and affordable power by planning, constructing, operating, and maintaining all generation, transmission, and distribution facilities in Nepal's power system, both interconnected and isolated.

Figure 1.2: Timeline of policy initiatives

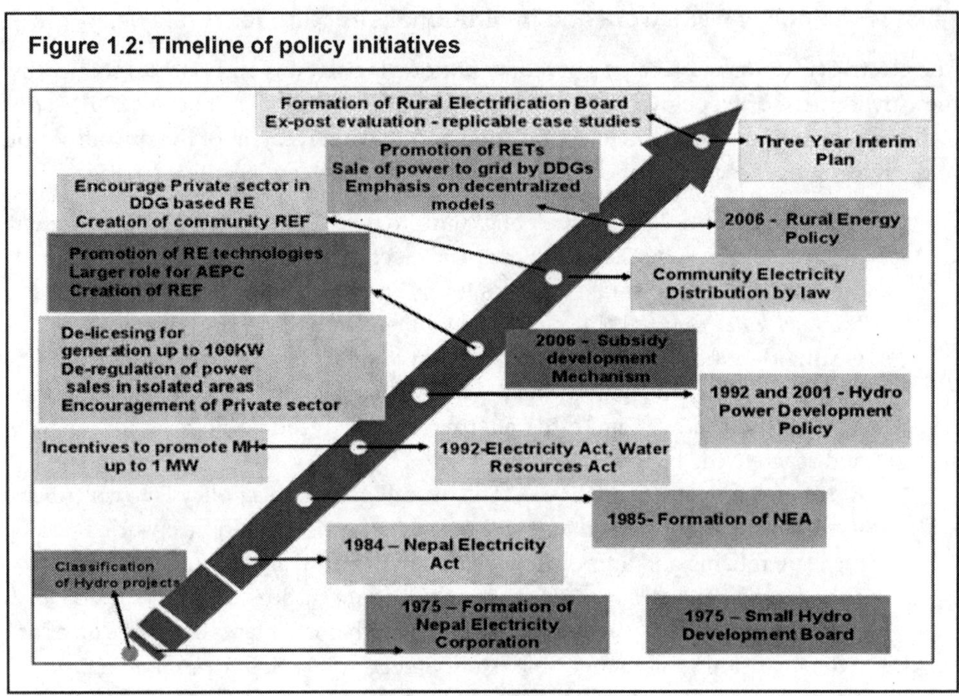

Source: Various policy documents of the Government of Nepal.

Figure 1.3: Institutional structure

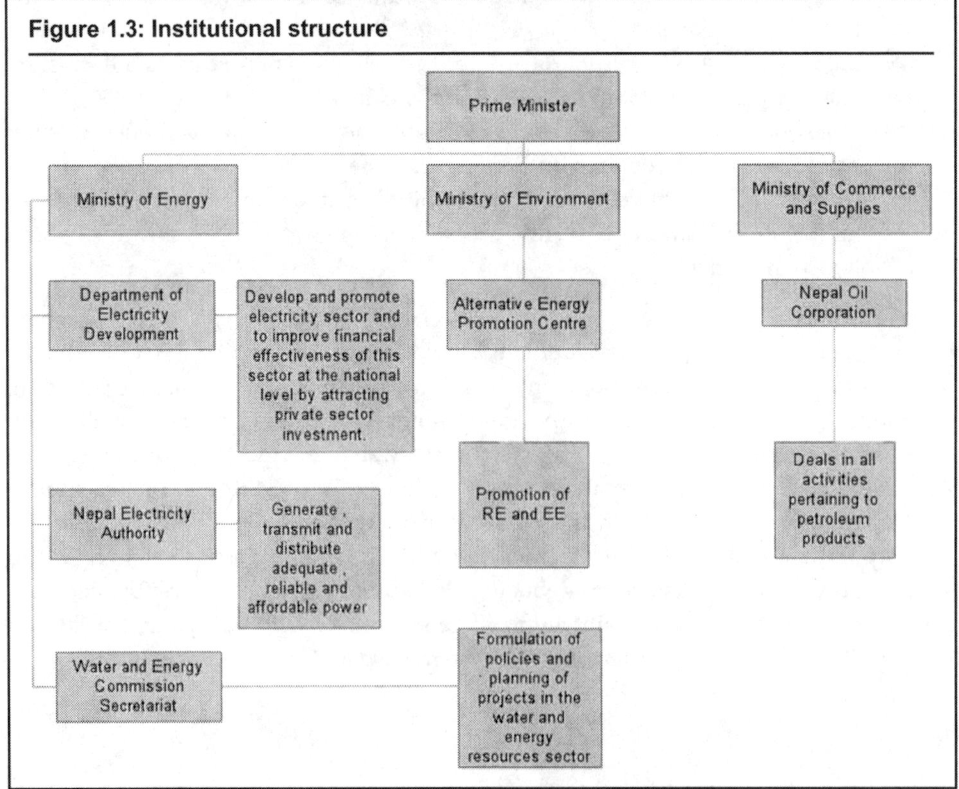

Source: Authors' elaboration.

The Small Hydropower and Rural Electrification Department in NEA is responsible for the construction, operation, and maintenance of small hydropower plants and the implementation of rural electrification programs in the remote hilly regions.

GoN adopted a policy in 2003 that calls for the sale of power to rural electricity consumer groups after setting up the requisite distribution infrastructure, to increase the coverage of electricity supply in the rural areas and to promote local participation for sustainable growth. Under this program, consumer associations, typically in the form of cooperatives and user groups, take the responsibility of managing, maintaining, and expanding the rural distribution of electricity.

Communities raise 20 percent of the investment cost for distribution system extension in their area and 80 percent of the fund is provided by the Government. The primary objectives of the program are to reduce the cost of distribution, make the distribution system effective, accelerate the pace of expansion of distribution lines in the rural areas, and check the pilferage of electricity.

However, NEA's financial performance has been declining significantly, and it is apparent that it is difficult for NEA to focus on rural electrification (figure 1.4). The investments in the rural electrification is a cost intensive sector due to the high cost of grid penetration, low connected load on the system, higher losses, and lower return.

Figure 1.4: NEA financial performance

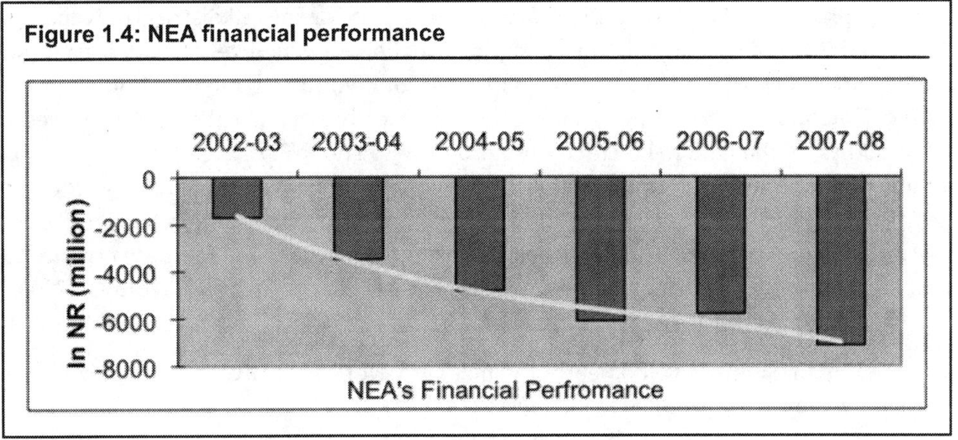

Source: NEA.

What are the program areas of AEPC?

AEPC, set up in 1996, has been tasked with the ambitious agenda of promoting rural electrification using renewable technologies (table 1.2). AEPC aims to energize rural households and kick-start economic activities in the project areas. AEPC has emerged as one of the world's leading proponents of community- and private-sector–led expansion of renewable energy technologies in rural areas. It is a Government institution falling under the oversight of Ministry of Environment and operating with semi-autonomous status. Its mandate is policy and plan formulation, resource mobilization, coordination, and quality assurance. The programs primarily relate to Micro/mini-hydro & Improved Water Mills, Solar energy (PV and thermal), Biogas, Improved cooking stoves, Wind energy, and Geothermal. Currently, it is supporting five programs funded by various donors.

Table 1.2: Major programs of AEPC

Programs	Major Support Areas	Donors
Energy Sector Assistance Program (ESAP)	Micro/mini-hydro, solar energy, biomass energy, institutional support	Danida, Norway
Rural Energy Development Program (REDP)	Micro/mini-hydro, community mobilization, environment, and sanitation	UNDP, World Bank
Renewable Energy Project (REP)	Community/institutional solar energy systems	EC
Biogas Support Program (BSP)	Biogas	DGIS-SNV, KfW
Improved Water Mill Support Program	Improved water mills	DGIS-SNV

Source: AEPC, 2009.

The oldest program is the REDP encompassing the micro-hydro facilities. The objective of the program is to enhance rural livelihoods and preserve the environment by better provisioning of energy through decentralized renewable energy technologies. The REDP Phase III, effective from 1 September 2007, is a joint program of the GoN, the United Nation Development Program (UNDP), and the World Bank (WB). REDP, which started with 5 districts (Phase I) in 1996, has now expanded into 40 districts of Nepal.

REDP is based on a holistic and integrated model of rural development. It uses the establishment of micro-hydro units as a catalyst for community mobilization and economic development. The program utilizes in an effective manner the existing government bodies on the national, district and local levels. The REDP project has evolved as an innovative award-winning model and has not only improved the quality of life but has also transformed the erstwhile divided community to stand and unite, take collective decision and prosper. The REDP works through a mechanism of self-governance to enable communities to design and manage their rural energy systems (Winrock, 2006). The organizing principles of REDP are six-fold:

- Organization development: Organize and work collectively to mobilize local resources for productive purposes.
- Capital mobilization: Encourage savings from community members to generate capital for development work.
- Skill enhancement: Enhance the skillset of rural people to use resources effectively and optimally compared to unskilled people.
- Technology promotion: Mainstream appropriate technology in the rural context to generate employment and replace dirty fuels.
- Environmental management: Conserve and manage water resources to enable sustainable operation of MH electricity.
- Vulnerable community empowerment: Empower women, ethnic groups, and vulnerable groups to participate in the design and operations of MH development.

REDP had slow growth during 2004–2005 but had a good year of development in 2006 (figure 1.5). Slow progress in 2004–2005 was largely due to insurgency in the country, which restrained the movement of machinery and equipment to the project areas. REDP's has established 272 micro-hydro plants with an installed capacity of 4,622 kW until April, 2010. Micro-hydro is the focal technology under the program; nonetheless, the program has created awareness of other rural energy systems as well.

Figure 1.5: Progress in the REDP

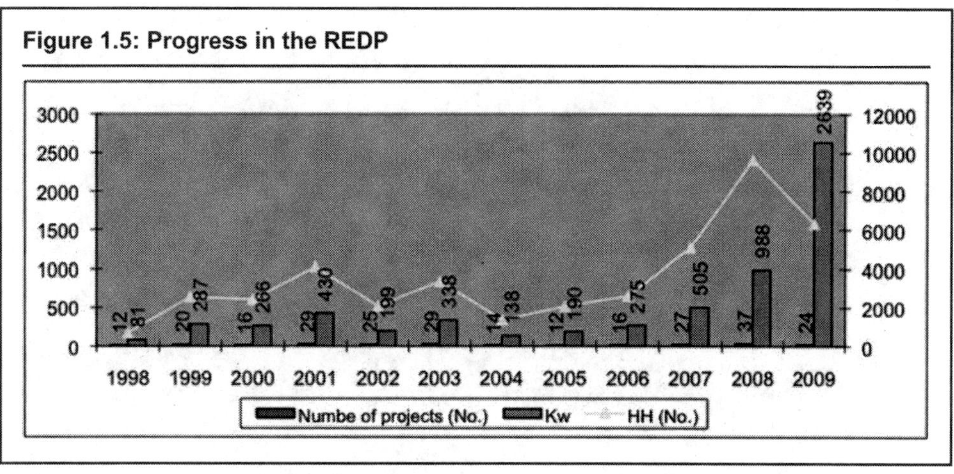

Source: Yearly achievements 1997–2009, AEPC.

REDP has created approximately 10,527 community organizations (COs) so far with an average of 25 members in each organization. On average, 28 COs are developed at one MH site and the average number of micro-enterprises created is 2.3 per site. Women make up about 50 percent of the COs' members. As on April 2010, cumulative investment of the COs amounts to Rs 110 million.

Aside from REDP, AEPC also supports programs such as Energy Sector Assistance Program (ESAP), Biogas Support Program (BSP), and Improved Water Mill (IWM). The renewable energy programs are supported by a number of donors. The ESAP began in 1999 and is funded by the Government of the Kingdom of Denmark (GKD), with a total budget of 154 million DKK. AEPC and NEA are the national agencies for the program. Physical outputs of ESAP I, including successive bridging periods and no-cost extensions, have reached more than 1.5 million people, who have benefited from improved cooking stoves and electric lights (solar home system and micro-hydro). In the year 2008-09, ESAP II reached to 35,098 HH (3.9 MW) under the Micro Hydro Program. The program disseminated 112160 SHSs, 52229 Metallic ICs, and commissioned 702 kW of hydro based schemes reaching to 6357 households under the Mini grid Support Program. The IWM Program was initiated in 2003 under the framework of the Renewable Energy Sector Support (RESS) Program of Government of Nepal and executed by AEPC. As on June 30, 2009, 5215 IWMs were installations that provide services to 250,000 HHs and 1,500,000 people (1 IWM provides services to 52 HHs).The BSP was initiated in July 1992 to develop and promote the use of Biogas, and laid down the objective to construct 7,000 Biogas plants. Based on the successful implementation of the first phase of the program, Phase II was started in July 1994 with the objective of constructing 13,000 Biogas plants. By March 2009, 189,237 biogas plants have been installed3. 96% biogas plants of the total installations are still functional.[4] Thus the program ensures that the plants installed are functional and people get the benefit of investment.

Notes

1. Economic Survey 2008-09, Ministry of Finance, 2009.
2. Nepal: Country Partnership Strategy, 2010–2014, A strategy for a country in transition.
3. Data against achievement under all program have been taken from the AEPC web site http://www.aepcnepal.org.
4. http://www.aepc.gov.np/index.php?option=com_content&view=article&id=86&Itemid=105.

CHAPTER 2

Objectives and Methodology of a Monitoring Framework Design for Renewable Energy

This chapter presents the objectives and methodologies of this study including the establishment of a baseline of welfare impacts of MH electrification in rural Nepal and development of a monitoring framework for AEPC.

The objective of this technical assistance (TA) to AEPC has been to enhance the ability of AEPC to manage and track the results of its investments. Given that AEPC's dependence on donor and Government funds is high, the need to develop a system that provides information on a wide range of technical, operational, and financial parameters is consequently high. The TA activity is also an unique effort to organize an evaluation system which at this point in time measures the impact of micro-hydro installations on rural livelihoods and to establish a monitoring system to continually measure the results of AEPC's renewable energy programs against the targets. With the support of the activity, AEPC is now equipped with not only the state-of-the-art monitoring system but also with a trained staff to sustainably manage and add to the system, as required.

This study encompassed four main components, developed in collaboration with AEPC:

(1) *Establishment of a monitoring framework:* The integrated monitoring framework encompasses all the programs of AEPC in micro-hydro, solar, and biomass programs. This workstream began with a diagnostic of AEPC's 'as-is' readiness and capacity, following which it identified the information needs that underpins monitoring and evaluation (M&E); developed a set of key performance indicators in a results chain; established a baseline against which future improvements will be tracked; prepared time-bound and realistic targets for the medium term (2009–2013) relative to reform efforts; and enhanced its ability to respond to information needs of internal and external stakeholders. This system has been developed in a consultative manner with AEPC staff to arrive at key indicators that best depict the operational, technical, and financial situation. The findings of this workstream are presented in Chapter 5.

(2) *Analysis of an evaluation of rural electrification with micro-hydro facilities:* There is significant anecdotal evidence of the positive impacts that access to electricity in rural areas in Nepal has on the lives of the poor, particularly on women. However, there is very little systematic quantification of these benefits through any well-designed evaluation framework. We fielded a 2,500 household survey and a 70 enterprise survey in early 2009 in rural Nepal to evaluate how micro-hydro-generated electrification has affected

welfare outcomes and quality of life across various dimensions. Using a counterfactual of non-micro-hydro–connected households, the analysis estimates to quantity the benefits accruing to micro-hydro connected households. A wide range of outcomes including quality of lighting, income generation, health, education, fertility, women's empowerment, and greenhouse gas (GHG) emissions reduction are considered. The findings of this evaluation are presented in Chapters 3 and 4.

(3) *Development of an MIS:* The workstream involved the implementation of the monitoring framework in the form of a MIS which was developed keeping in mind that the data should be presented in a timely and concise manner to the relevant internal and external target audience and at different levels within AEPC. This activity proceeded with defining reporting requirements and formats for standard reports, which include the data to be collected, the source of data, the frequency of data collection, and the parties responsible for collecting, analyzing, reporting, and using the data. The findings of this workstream are presented in Chapter 5.

To enable AEPC staff to use this system, two rigorous training sessions were held and a detailed operations manual was developed. The study has also recommended the expansion of the M&E cell in AEPC and specified the technical specifications of servers that can sustainably support this new MIS.

(4) *Development of a diagnostic and a forward-looking business plan for AEPC:* AEPC, as a nodal organization for renewable energy development, has to chart a way forward and establish its priorities based on competencies and comparative advantage. This business plan has to take into account a number of parallel initiative and policy frameworks, namely the Strategic Organization Development Plan (SODP), a 3-year interim plan, and a rural energy policy. The findings of this workstream are presented in Chapter 6 and will form an input to AEPC's Strategic Organization Development Plan that is currently being updated.

Why is monitoring important for AEPC?

Monitoring the performance and assessing the impact of renewable energy programs is primarily the responsibility of AEPC. AEPC as the nodal agency is facing increasing demand to measure the performance of its renewable energy programs. There is an urgent need to better respond to the information needs of its internal and external stakeholders. AEPC is required to report to the GoN on the progress and effectiveness of renewable energy programs and to demonstrate results to the funding agencies against the money spent. Also, there is a general need to provide management and staff with real-time feedback on various aspects of the performance of specific programs assessed, and to better understand the intermediate and final impact of its interventions to improve design and implementation of projects.

In response, AEPC has strengthened its monitoring arrangements and reporting processes. AEPC has made an impressive beginning in measuring certain performance indicators for various programs. There is routine collection of information on project implementation and reporting on inputs and provision of outputs. A brief overview of the existing M&E functions of various programs in AEPC is presented in Annex 1. With the help of a third-party agency, AEPC has also been evaluating the outcomes of various programs occasionally. There is also significant anecdotal evidence of the positive impact of these programs on beneficiaries. Moreover, the presence of dedicated task teams with strong commitment to monitoring the achievement of program objectives merits highlighting.

These initiatives are necessary, but they are not yet sufficient. A lot of information is also reported through anecdotal evidence or through qualitative indicators. There is very little systematic quantification of the benefits through a well-designed M&E framework. The stakeholders are not solely interested in the program's activities and outputs but on how these interventions enhance people's lives. Concerns have grown about lack of routine reporting on the progress and effectiveness of its programs.

How was the monitoring and evaluation (M&E) framework developed?

The primary objective of this study has been to develop a robust yet simple M&E framework for all the programs of AEPC that is focused on the needs of the decision-makers, as well as the interests of the relevant stakeholders. The monitoring system has been developed thorough extensive involvement of AEPC officials during the conceptualization, design, testing, and implementation of the system. The M&E system builds on the existing M&E systems of AEPC and ensures linkages to the existing systems for various programs.

A well-designed M&E framework supported by a robust MIS is a useful tool that would enable AEPC management to track progress and demonstrate the impact of its programs. The M&E system can help support the following objectives:

- Support GoN in policy formulation and program development through analysis of the effectiveness of the program.
- Support AEPC management in strategy formulation and resource allocation.
- Promote greater transparency and accountability by demonstrating results of AEPC-supported programs to the internal and external stakeholders.
- Promote organizational learning through providing continuous feedback in managing the program.
- Facilitate benchmarking among different regions and districts.

What are the attributes of the monitoring and evaluation framework?

Focus on all aspects in the results chain: The focus has been to develop performance indicators across the entire causal chain from project intervention to on-the-ground impacts. The focus has been to identify the information needs that underpin M&E based on a clear understanding of the causal chain through which the project interventions will lead to the desired outputs and outcomes. The aim has been to create the results chain from resource and community mobilization to installation of systems to improved provision of services, to their actual use and impact on lives. The focus is not only the activities undertaken and the outputs but also the impact on the beneficiaries to gain a better perspective of the impact of the interventions and to support future planning processes and decision-making. The M&E framework for integrated renewable energy projects is presented in Figure 2.1.

The indicators have been classified in the following heads:

- *Inputs:* This includes financial, human, and other technical resources mobilized to support activities undertaken by AEPC. Examples include funds utilized, value of land, number of awareness-generation programs for community mobilization, etc.
- *Outputs:* This includes the interventions by AEPC in terms of technical aspects such as systems installed, number of households connected; capacity building such as number of staff trained; and community participation such as number of community organizations, etc.
- *Outcomes:* The outcomes measure the uptake, adoption, and use of services by the project beneficiaries. In the context of AEPC, these are assessed on various parameters including access, affordability, operating efficiency, reliability, and financial sustainability.
- *Impacts:* These include long-term outcomes near or at the end of the results chain such as the effect of provision of electricity on employment, health, education, and income. Since this information needs to be collected at a household level, it is suggested to be collected every 2–3 years based on a survey.

Key Performance Indicators (Input, Output, Outcome). The baseline data for the inputs, outputs, and outcome KPIs were gathered with the objective to establish a baseline against which the future improvements will be tracked on a regular basis. Particular attention was paid to ensure that the M&E system supplements and links to the existing M&E system of the client and takes into consideration the capacity of AEPC officials.

The KPIs have been selected to facilitate informed decision-making. The indicators are monitored throughout implementation to assess progress toward program objectives. These standard indicators have been identified for each of the ongoing programs in AEPC. The KPIs have been designed based on the following principles:

Incorporation of key principles: As reflected in various policy and plan documents of the Government of Nepal, M&E has increasingly been assigned more importance in recent policy and plan documents. The indicators incorporate various aspects of accessibility, reliability, affordability, and community participation, which have been reflected as key objectives in various policy and planning documents of the Government of Nepal.

Alignment with AEPC's strategic priorities: The KPIs have been designed to reflect the key objectives and strategic priorities of AEPC as reflected in its Mission statement: *"Mainstream renewable energy resource through increased access, knowledge & adaptability contributing for the improved living conditions of people in Nepal."*

Alignment with key objectives of the programs: The indicators have been developed to measure the key objectives of the programs managed by AEPC and the reforms it plans to undertake in the medium term. For instance, in the case of REDP, indicators have been developed to enable the management to assess impact on its key principles, i.e., organizational development, skill enhancement, capital formation, technology promotion, community empowerment, and environmental management. In the case of

Figure 2.1: M&E framework for assessing rural energy impacts

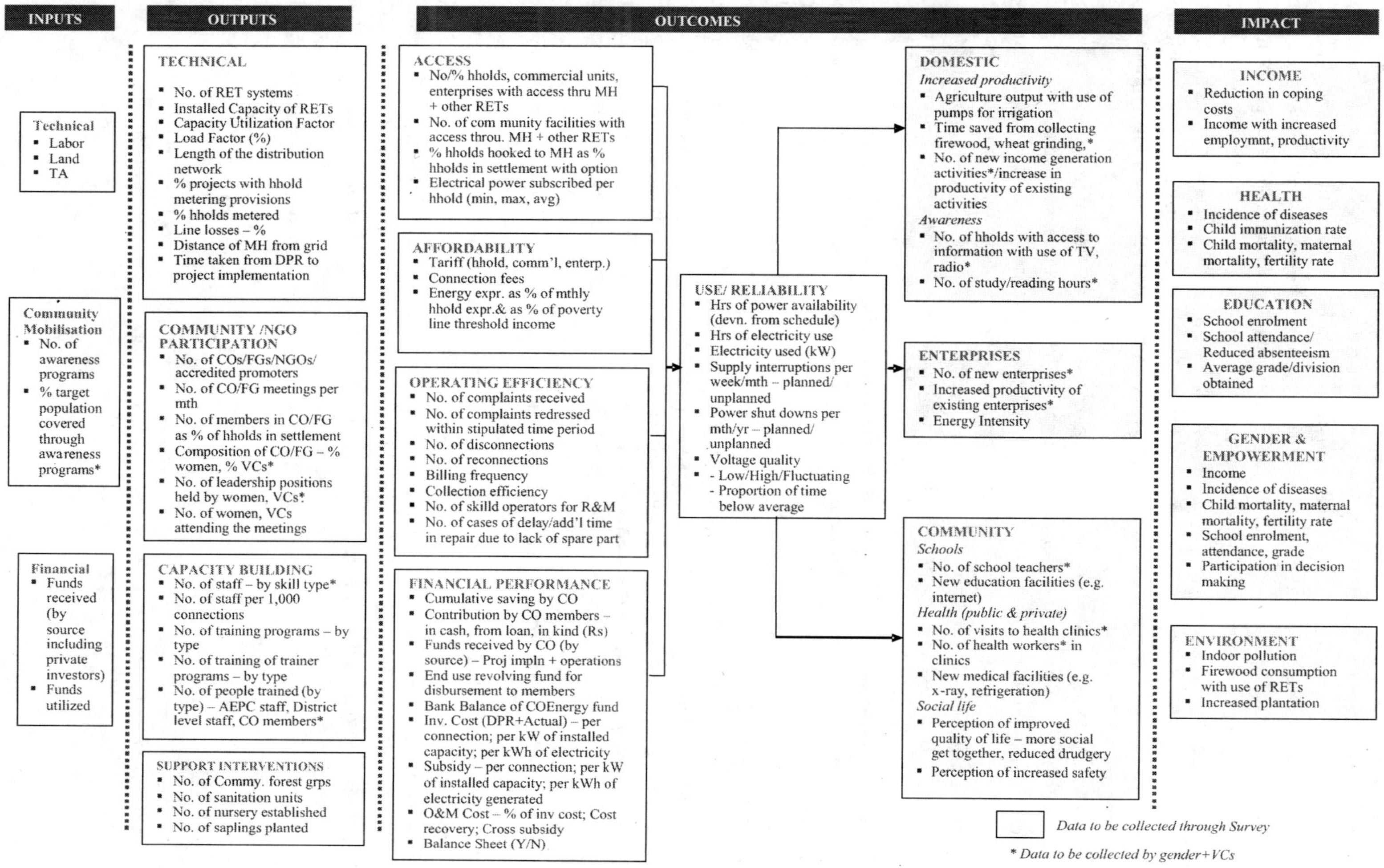

INPUTS

Technical
- Labor
- Land
- TA

Community Mobilisation
- No. of awareness programs
- % target population covered through awareness programs*

Financial
- Funds received (by source including private investors)
- Funds utilized

OUTPUTS

TECHNICAL
- No. of RET systems
- Installed Capacity of RETs
- Capacity Utilization Factor
- Load Factor (%)
- Length of the distribution network
- % projects with hhold metering provisions
- % hholds metered
- Line losses – %
- Distance of MH from grid
- Time taken from DPR to project implementation

COMMUNITY /NGO PARTICIPATION
- No. of COs/FGs/NGOs/ accredited promoters
- No. of CO/FG meetings per mth
- No. of members in CO/FG as % of hholds in settlement
- Composition of CO/FG – % women, % VCs*
- No. of leadership positions held by women, VCs*
- No. of women, VCs attending the meetings

CAPACITY BUILDING
- No. of staff – by skill type*
- No. of staff per 1,000 connections
- No. of training programs – by type
- No. of training of trainer programs – by type
- No. of people trained (by type) – AEPC staff, District level staff, CO members*

SUPPORT INTERVENTIONS
- No. of Commy. forest grps
- No. of sanitation units
- No. of nursery established
- No. of saplings planted

OUTCOMES

ACCESS
- No/% hholds, commercial units, enterprises with access thru MH + other RETs
- No. of com munity facilities with access throu. MH + other RETs
- % hholds hooked to MH as % hholds in settlement with option
- Electrical power subscribed per hhold (min, max, avg)

AFFORDABILITY
- Tariff (hhold, comm'l, enterp.)
- Connection fees
- Energy expr. as % of mthly hhold expr.& as % of poverty line threshold income

OPERATING EFFICIENCY
- No. of complaints received
- No. of complaints redressed within stipulated time period
- No. of disconnections
- No. of reconnections
- Billing frequency
- Collection efficiency
- No. of skilld operators for R&M
- No. of cases of delay/add'l time in repair due to lack of spare part

FINANCIAL PERFORMANCE
- Cumulative saving by CO
- Contribution by CO members – in cash, from loan, in kind (Rs)
- Funds received by CO (by source) – Proj impln + operations
- End use revolving fund for disbursement to members
- Bank Balance of COEnergy fund
- Inv. Cost (DPR+Actual) – per connection; per kW of installed capacity; per kWh of electricity
- Subsidy – per connection; per kW of installed capacity; per kWh of electricity generated
- O&M Cost – % of inv cost; Cost recovery; Cross subsidy
- Balance Sheet (Y/N)

USE/ RELIABILITY
- Hrs of power availability (devn. from schedule)
- Hrs of electricity use
- Electricity used (kW)
- Supply interruptions per week/mth – planned/ unplanned
- Power shut downs per mth/yr – planned/ unplanned
- Voltage quality
- - Low/High/Fluctuating
  - Proportion of time below average

DOMESTIC

Increased productivity
- Agriculture output with use of pumps for irrigation
- Time saved from collecting firewood, wheat grinding,*
- No. of new income generation activities*/increase in productivity of existing activities

Awareness
- No. of hholds with access to information with use of TV, radio*
- No. of study/reading hours*

ENTERPRISES
- No. of new enterprises*
- Increased productivity of existing enterprises*
- Energy Intensity

COMMUNITY

Schools
- No. of school teachers*
- New education facilities (e.g. internet)

Health (public & private)
- No. of visits to health clinics*
- No. of health workers* in clinics
- New medical facilities (e.g. x-ray, refrigeration)

Social life
- Perception of improved quality of life – more social get together, reduced drudgery
- Perception of increased safety

IMPACT

INCOME
- Reduction in coping costs
- Income with increased employmnt, productivity

HEALTH
- Incidence of diseases
- Child immunization rate
- Child mortality, maternal mortality, fertility rate

EDUCATION
- School enrolment
- School attendance/ Reduced absenteeism
- Average grade/division obtained

GENDER & EMPOWERMENT
- Income
- Incidence of diseases
- Child mortality, maternal mortality, fertility rate
- School enrolment, attendance, grade
- Participation in decision making

ENVIRONMENT
- Indoor pollution
- Firewood consumption with use of RETs
- Increased plantation

Data to be collected through Survey

** Data to be collected by gender+VCs*

Source: Authors.

ESAP, they help measure various aspects of its objectives in terms of improving the living conditions of the rural population by enhancing their access and affordability to rural energy solutions that are environmentally friendly and that address social justice. For IWM and Biogas programs, the focus has been to address poverty, social inclusion, and regional balance issues by further mainstreaming these renewable energy solutions.

Key performance indicators (Impact). Impact indicators were developed as the final outcomes of the micro-hydro interventions on household welfare and enterprise productivity.[1] The baseline values of these indicators were established using a primary survey in rural Nepal. AEPC can repeat these surveys every few years to evaluate and quantify the final benefits of its programs

The primary rural household and enterprise survey was conducted jointly by the World Bank and AEPC in Nepal during early 2009. It also uses a wide range of publicly available district-level information to complement the survey data. Furthermore, it draws on reports and presentations prepared by AEPC and the World Bank. For the household survey, 2,500 households were selected, of which 1,500 are treated (MH connected), 571 are control from MH locations, and 429 are control from non-MH (control) locations. Seventy microenterprises were selected from the same locations that the households were selected from, and their selection followed similar criteria.

The survey contains detailed questions on household expenditure patterns, and welfare outcomes including income, education, health, housing, consumption, assets, farm and non-farm production, and women's empowerment. In addition, there is a module on the household's energy use that covers details on the type of energy the household uses, and their quantity, cost, appliances supported, and so on. To complement household information, questions were also asked at the community level to obtain information on the community prices of different energy sources and consumer goods, wages of males, females, and children, and infrastructure information. Besides household and location questionnaires, an enterprise questionnaire was fielded to investigate the energy use pattern of local microenterprises and benefits of MH on the products and services produced by those enterprises. The sample selection is described in detail in Annex 3.

The welfare outcomes of the MH households is compared with that of control households, and the difference in the means of outcomes between the two types of households provides an estimate of the electrification benefits accrued by MH households. However, benefit estimated this way may be biased, because MH and non-MH households may vary fundamentally in other characteristics, not just in MH connectivity, and it is difficult to determine whether the estimated benefits are due to MH access or to other characteristics. This problem has been addressed by using a matching technique, in particular, a propensity score matching (PSM) technique, which is the most common among matching techniques.

Notes

1 The evaluation was carried out only for the micro-hydro projects as this is the oldest program (REDP) and not enough time has passed for other programs to gauge the final benefits.

CHAPTER 3

Coverage and Attributes of Micro-Hydro for Households and Enterprises

This chapter discusses the energy use patterns, spending on and quality of service of MH and non-MH households from the World Bank-AEPC Survey, 2009 carried out as part of this study.[1]

Rural electricity coverage in Nepal, a country featuring many difficult terrains, has lagged significantly behind urban areas. The initiatives to expand connections and bring power to rural households have focused primarily on off-grid sources such as micro-hydro and solar. Micro-hydro (MH), specifically, has carved an important niche among decentralized energy options in rural Nepal. REDP, pioneering the community designed and managed option, has successfully brought electricity to many villages in rural Nepal.

What is the pattern of rural households' energy use?

Households with better human capital endowment (education for example), and physical endowment (asset) are more likely to have MH than those that are worse off. For example, the average education level of adult females in a household that has MH is 3.73 years, compared to 2.90 years of education of adult females in a non-MH household (table 3.1). Adult males in MH households also have higher education than those in non-MH households, however the difference is not statistically significant. Households with more land and non-land assets are more likely to have MH, so are those with better hygiene.

Table 3.1: HH characteristics by MH access

HH characteristics	MH users	Nonusers of MH	t-statistic (MH users–Nonusers)
Max. education of adult males (years)	5.28	4.74	1.46
Max. education of adult females (years)	3.73	2.90	2.49**
Land asset (acre)	1.50	1.28	1.89*
Non-land asset (Rs.)	178,966.4	125,515.7	4.11**
HH has running water	0.92	0.77	4.42**
HH has pit latrine or better	0.50	0.34	4.15**
Observations	1,499	998	

Note: * and ** refer to a significance level of 10 percent and 5 percent respectively.
Source: World Bank-AEPC survey 2009.

Biomass energy, fuel wood in particular, is by far the most dominant form of energy for rural households regardless of their MH connectivity. Overall, almost 100 percent of the households use fuel wood, and very few of them use dung and straw. Households get their biomass in three ways: their own production (for example, fuel wood from own trees, dung from own cattle), collection from locally available free sources (mostly fuel wood), and some from purchase.

Kerosene is used widely among the commercial sources. MH users use it, understandably, less than nonusers of MH. Since the primary use of both MH and kerosene is lighting, once the households get access to MH, they replace kerosene with MH as their primary lighting source. However, kerosene use is still quite high among the MH users

Table 3.2: Energy use pattern of sample households across regions (%)

Energy sources	MH users	Nonusers of MH	Whole sample
Biomass			
Fuel wood	100.0	99.9	99.9
Dung	1.0	0.1	0.1
Straw	4.7	0.7	0.9
Commercial			
Kerosene	74.9	85.9	85.2
LPG	0.5	0.3	0.3
Charcoal	0.3	0.3	0.3
Coal	0.6	0.2	0.2
RET			
PV/SHS	0	5.3	5.0
Biogas	6.2	5.9	5.9
Micro-hydro (MH)	100.0	-	6.1
Misc. small sources			
Candle	16.2	11.1	11.4
Dry cell	59.5	79.9	78.7
Observations	1,499	998	2,497

Source: World Bank-AEPC survey 2009.

(almost 75 percent), probably because they use it as a backup source for lighting. On the other hand, over 85 percent of the nonusers of MH use kerosene. Rural households in Nepal also use liquid petroleum gas (LPG), charcoal, and coal, although use of these fuels is very limited.

Renewable energy forms a small proportion of rural households' energy use. Six percent of the households have access to MH, and among the nonusers of MH a little over 5 percent use PV/SHS (table 3.2). Both MH users and nonusers are dependent on biogas (roughly 6 percent). Besides these sources, miscellaneous small energy sources, for example, candle and dry cell battery, are also used by the rural households. Use of candle is 16 percent among the MH users and 11 percent among the nonusers of MH. About 60 percent of the MH users use dry cell, while its use among the nonusers of MH is almost 80 percent. Dry cell finds most of its use in small appliances (transistors, flash lights, calculators) which are not powered by MH electricity. Although a significant number of households use candle and dry cell, the energy content of these sources is very small, and therefore, their contribution to overall energy consumption of the households is insignificant.

There is no significant difference in biomass use between MH users and nonusers. All households, regardless of their MH connectivity, use fuel wood for cooking, which is the primary use of biomass (table 3.3). Kerosene is also used for cooking. Ten percent of the MH users and about 4 percent of the nonusers of MH use it for cooking. However, distinction between MH users and nonusers is more obvious in their use of kerosene for lighting. Use of LPG is very low in rural Nepal, and it is used mostly for cooking. About 6 percent of the households use biogas for cooking and only about 1 percent use it for lighting. About 59 percent of the MH users and 78 percent of the non-MH users employ dry cell batteries for lighting—most likely in flashlights. Dry cell batteries are also used in small appliances by 16 percent of the non-MH households. All users of MH use it for lighting. Almost 10 percent of them also use it in appliances, mostly TV/VCR (or DVD player), and to some extent electric chargers. Finally, households rarely use energy for home business and the two energy sources that are used for that purpose are coal and charcoal. Overall, less than one percent of the households use energy for home business.

Table 3.3: Share of HHs using different energy sources for various purposes (%)

Energy sources	Cooking	Lighting	Heating	Appliance use	Home business
MH users					
Biomass					
Fuel wood	100.0	0.0	10.0	0.0	0.0
Dung	1.0	0.0	0.0	0.0	0.0
Straw	4.7	0.0	0.0	0.0	0.0
Commercial					
Kerosene	10.0	72.6	0.0	0.0	0.0
LPG	0.5	0.0	0.0	0.0	0
Charcoal	0.0	0.0	0.0	0.0	0.3
Coal	0.0	0.0	0.0	0.0	0.6
RET					
PV/SHS	0	0	0	0	0
Biogas	6.2	0.0	0.0	0.0	0.0
Micro-hydro (MH)	0.5	100.0	0.0	9.5	0.0
Misc. small sources					
Candle	0.0	16.2	0.0	0.0	0.0
Dry cell	0.0	58.8	0.0	4.1	0.0
Non users of MH					
Biomass					
Fuel wood	100.0	0.0	21.9	0.0	0.0
Dung	0.1	0.0	0.0	0.0	0.0
Straw	0.7	0.0	0.0	0.0	0.0
Commercial					
Kerosene	3.9	84.8	0.0	0.0	0.0
LPG	0.2	0.04	0.0	0.0	0
Charcoal	0.0	0.0	0.0	0.0	0.3
Coal	0.0	0.0	0.0	0.0	0.2
RET					
PV/SHS	0.0	5.3	0.0	0.0	0.0
Biogas	5.9	1.3	0.0	0.0	0.0
Micro-hydro (MH)	-	-	-	-	-
Misc. small sources					
Candle	0.0	11.1	0.0	0.0	0.0
Dry cell	0.0	78.0	0.0	17.2	0.0
Whole sample					
Biomass					
Fuel wood	100.0	0.0	21.2	0.0	0.0
Dung	0.1	0.0	0.0	0.0	0.0
Straw	0.9	0.0	0.0	0.0	0.0
Commercial					
Kerosene	4.3	84.1	0.0	0.0	0.0
LPG	0.3	0.04	0.0	0.0	0
Charcoal	0.0	0.0	0.0	0.0	0.3
Coal	0.0	0.0	0.0	0.0	0.2
RET					
PV/SHS	0.0	5.0	0.0	0.0	0.0
Biogas	5.9	1.3	0.0	0.0	0.0
Micro-hydro (MH)	0.03	6.1	0.0	0.6	0.0
Misc. small sources					
Candle	0.0	11.4	0.0	0.0	0.0
Dry cell	0.0	76.8	0.0	16.4	0.0

Source: World Bank-AEPC survey 2009.

Females spend more time than males in fuel collection (table 3.4). However, their difference is statistically significant only for nonusers of MH. Biomass collection time for MH users is more than that for nonusers of MH. That may be because MH households, being better-off, just cook more and so consume more biomass. Overall, household members spend roughly about 6 hours per month in collecting biomass.

Table 3.4: Household's biomass collection time (hours/month)

Collection variables	MH users	Nonusers of MH	Whole sample	t-stats
Male hours	3.01	2.76	2.91	1.73*
Female hours	3.11	2.96	3.05	1.10
Child hours	0.32	0.31	0.31	0.11
t-stats	t_{MF}=1.14, t_{FC}=28.3**, t_{MC}=24.2**	t_{MF}=-1.91*, t_{FC}=25.3**, t_{MC}=26.1**	2,497	2,497
Observations	1,499	998	2,497	2,497

Note: * and ** refer to a significance level of 10 percent and 5 percent respectively.
Source: World Bank-AEPC survey 2009.

What is the pattern of rural households' energy consumption?

Households' energy need is mostly met by biomass (table 3.5). Nepal is rich in forest resources and these resources are used heavily to satisfy fuel needs and for development activities. According to a 2005 study, about 44,000 hectares of forest are decimated annually in Nepal, while only 4,000 hectares are replenished (CRT/N, 2005). MH users consume slightly more biomass energy (80 kgOE/month) than the nonusers of MH (76 kgOE/month). High reliance on biomass is an indication of the inadequate development of alternate energy sources, besides being a deforestation threat (since most of biomass comes from fuel wood).

Once again, kerosene consumption distinguishes MH users from the nonusers. MH users consume 0.96 kgOE of energy from kerosene, which is less than half of what the non-MH households consume. That is a saving of more than one liter of kerosene per month for MH users. A good portion of the kerosene that MH users consume is also used for cooking, while non-MH households use kerosene mostly for lighting.

The average electricity consumption of MH users is only about 1.9 kgOE/month (little over 22 kWh/month), which is roughly equivalent to 4 hours of daily use of three 60-watt bulbs. The consumption of MH electricity in rural Nepal is limited by a capacity ceiling that is adequate for just basic lighting and limited use of low-wattage appliances or gadgets, and that is why simultaneous use of all of them is often not possible. The capacity limitation is imposed because MH plants are themselves of low capacity (about 5 to 100 kW) and imposing a capacity limitation is the only way to ensure MH to more households.[2]

Table 3.5: Energy consumption of sample households by MH access (kgOE/month)

Energy sources	MH users	Nonusers of MH	Whole sample
Biomass	80.30	76.24	76.49
Kerosene	0.96	1.96	1.89
LPG	0.08	0.04	0.05
PV/SHS	0.0	0.02	0.019
Biogas	0.52	0.76	0.73
Micro-hydro (MH)	1.90	0	0.12
Other sources[1]	1.07	0.75	0.77
Total	84.80	79.77	80.06
Observations	1,499	998	2,497

[1] Other sources include charcoal, coal, candle, and dry cell.
Source: World Bank-AEPC survey 2009.

Average household consumption of energy from LPG is quite low at 0.05 kgOE/month, mostly because very few households use it. Consumption of solar energy (PV/SHS) is also very low, which is used, not surprisingly, for lighting only by non-MH households. Other miscellaneous small sources (coal, charcoal, candle, and dry cell) add about one kgOE to the energy consumed by MH users and 0.75 kgOE to that consumed by non-MH households. Overall, MH households consume a little bit more energy (85 kgOE/month) than the non-MH households (80 kgOE/month).

How much do rural households spend on energy?

Rural households in Nepal spend Rs. 875 per month on energy (table 3.6), and expenditure on biomass is by far the highest. At Rs. 584, biomass expenditure constitutes 67 percent of the total energy expenditure.[3] The next big expenditure item is kerosene. Non-MH users spend almost Rs. 200 per month on kerosene (about 23 percent of their total energy expenditure), while MH households spend about Rs. 92 on kerosene. Expenditure on MH is Rs. 62 per month. Given this expenditure, households' MH electricity consumption is not optimal, because households pay more than what they consume. There is no metering system for MH, and households are charged a flat amount based on their capacity ceiling, which limits the number of appliances they can use. Since households are not charged for the duration of use, they pay proportionately more if they do not use their appliances long enough.

Table 3.6: Energy expenditure of sample households by MH access (Rs./month)

Energy sources	MH users	Nonusers of MH	Whole sample
Biomass	624.05	581.98	584.54
Kerosene	91.60	199.13	192.56
LPG	6.29	3.90	4.05
PV/SHS	-	4.97	4.67
Biogas	5.71	8.55	8.37
Micro-hydro (MH)	62.46	-	3.82
Other sources[1]	47.86	79.31	77.39
Total	838.08	877.84	875.41
Observations	1,499	998	2,497

[1] Other sources include charcoal, coal, candle, and dry cell.
Source: World Bank-AEPC survey 2009.

Non-MH households spend about 40 Rs. more than MH households but consume slightly less energy than MH users. What it means is that non-MH households pay more for their energy consumption than MH households.

Rural households in Nepal on an average pay close to Rs. 11 for one kgOE. Based on the estimations of households' payment for energy consumed from different sources (Rs./kgOE), MH households pay slightly less (9.75 Rs./kgOE) than their counterpart non-MH households (10.80 Rs./kgOE) (table 3.7). Although the price of energy from MH is more than that from few other sources (for example, biomass), it is in reality much cheaper because one kgOE of electricity can provide a much wider range of energy services than one kgOE of biomass or any other sources.

Table 3.7: Price paid for energy by sample households by MH access (Rs./kgOE)

Energy sources	MH users	Nonusers of MH	Whole sample
Biomass	7.77	7.63	7.64
Kerosene	95.55	101.81	101.63
LPG	81.42	89.69	88.83
PV/SHS	-	245.24	245.24
Biogas	10.97	11.10	11.08
Micro-hydro (MH)	32.88	-	32.88
Other sources[1]	44.55	105.71	100.47
Total	9.75	10.80	10.73
Observations	1,499	998	2,497

[1] Other sources include charcoal, coal, candle, and dry cell.
Source: World Bank-AEPC survey 2009.

Household energy expenditure rises with rising income, but at a slower rate. But the share of energy expenditure as a percentage of total income goes down steadily as household income goes up, which is also expected. For example, households in the first quintile, who have an average income of about Rs. 3,353, spend 22 percent of their income on energy.[4] On the other hand, wealthiest households (income quintile 5) spend just 6.4 percent of their income on energy. Overall, rural households in Nepal spend 11 percent of their income on energy (table 3.8).

Table 3.8: Household's energy expenditure by income quintile

Income/energy variables	1st	2nd	3rd	4th	5th	All households
Income (Rs./month)	3,352.9	5,590.9	6,922.8	9,294.5	16,387.9	7,679.3
Energy expenditure (Rs./month)	737.6	829.3	860.6	972.5	1,055.4	875.4
Energy expenditure as a percentage in income	22.0	14.8	12.4	10.5	6.4	11.4
Observations	500	499	500	499	499	2,497

Source: World Bank-AEPC survey 2009.

What is the volume of rural households' MH use?

A majority of MH consumers have been served for about 6 to 8 years (table 3.9). On an average, households have had access to MH for 6.7 years. Consumption of electricity does not show a consistent pattern as the duration goes up. Households that have MH access for 8–10 years consume the most (31 kWh/month), followed by those who are relatively new, with average consumption of 29 kWh/month.

Table 3.9: HH MH energy consumption by the duration of their MH access

MH variables	<=2 years	2-4 years	4-6 years	6-8 years	8-10 years	10+ years	Mean for all HHs
% of HHs	4.6	12.9	8.0	43.9	20.9	9.7	6.7
MH consumption (kWh/month)	11.7	29.1	20.4	17.7	31.4	6.9	22.05

Note: Last column shows average duration and consumption for all MH users.
Source: World Bank-AEPC survey 2009.

One important reason for connecting to MH is children's education. An overwhelming 92 percent of users concur with this view, followed by 61 percent stating that the low price of electricity is the reason for their MH connectivity.

High-income households consume more electricity than their lower income counterparts (table 3.10). However, the trend in average price paid (Rs/kWh) is not very consistent, although it goes down as the income goes up. Therefore, high-income households not only consume more electricity (either by using more appliances or using them for longer duration) than low-income households, they pay a lower price. Overall, highest income households consume almost twice as much electricity as the lowest income households do, while paying only 32 percent more.

Table 3.10: Household's MH energy consumption by income quintile

MH use variables	1st	2nd	3rd	4th	5th	All households
Electricity consumption (kWh/month)	16.29	19.08	18.39	23.44	32.19	22.05
Electricity expenditure (Rs./month)	54.15	58.28	58.47	69.51	71.48	62.46
Price paid (Rs./kWh)	3.32	3.05	3.18	2.97	2.19	2.83
Observations	300	300	300	300	299	1,499

Note: The sample is restricted to MH users only.
Source: World Bank-AEPC survey 2009.

MH electricity consumption varies by region (figure 3.1). Households in the western region (which have the districts of Tanhun, Myagdi, Parvat, and Baglung) consume the highest quantity of electricity (almost 29 kWh/month), while those in the far west region (districts Achham, Dadeldhura, and Baitadi) consume the least (about 11 kWh/month).

Figure 3.1: Regional variation in MH consumption

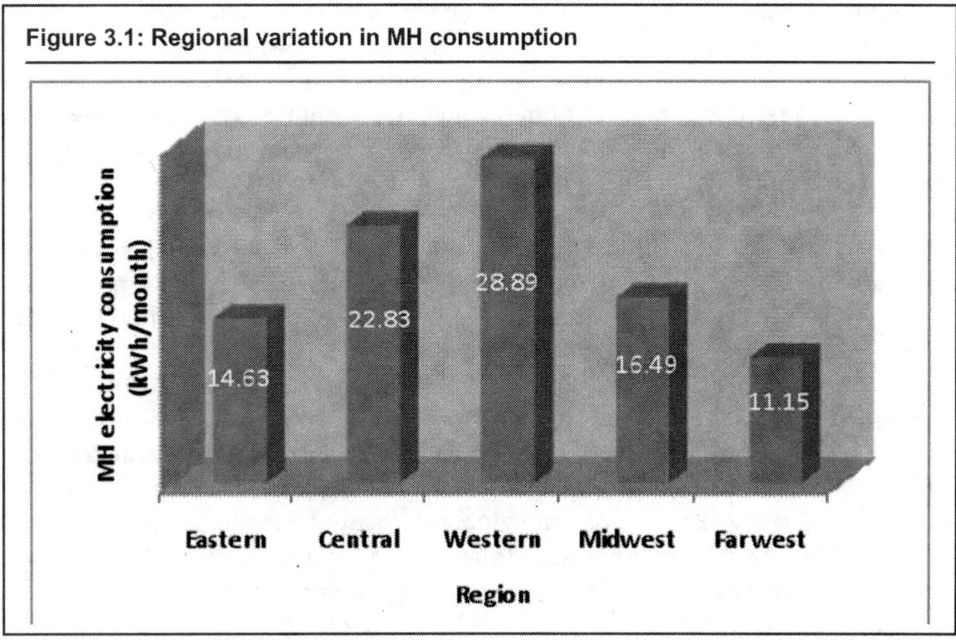

Source: World Bank-AEPC survey 2009.

While the primary use of MH electricity is lighting, some households use it for appliances (table 3.11). Among the light bulbs, incandescent bulbs are the most common (among which 24w ones are owned by most households). Households use more incandescent bulbs as their income goes up—while lowest quintile households own on an average 2.4 bulbs, the highest quintile households own 3.6 bulbs. On an average, MH households own about 3 bulbs. Besides incandescent bulbs, fluorescent and CFL bulbs are also used by a small share of MH households. These bulbs are more expensive than incandescent bulbs but they are more energy efficient and durable.

Table 3.11: MH household's ownership of different appliances by income quintile

Appliance	1st	2nd	3rd	4th	5th	All households
No. of incandescent bulbs	2.44	2.61	2.73	3.11	3.62	2.92
No. of fluorescent bulbs	0.58	0.39	0.71	0.80	0.71	0.64
No. of CFL bulbs	0.10	0.17	0.24	0.17	0.16	0.16
% TV owners	4.01	2.94	3.46	10.87	15.61	7.58
% radio owners	1.70	2.57	2.75	3.79	7.79	3.80
% DVD/VCD player owners	1.28	1.20	0.76	1.20	1.34	1.16
Observations	300	300	300	300	299	1,499

Note: The sample is restricted to MH users only.
Source: World Bank-AEPC survey 2009.

Among appliances, TV is the most common among the MH households. About 7.5 percent of MH households own TVs, and as expected, TV ownership goes up consistently with rising income. While only 4 percent of the lowest income households own a TV, the ownership is almost four times among the wealthiest group of households. Households also own radios and DVD/VCD players. The ownership of radio follows the same pattern as that of TV ownership as a household's income goes up, while ownership of DVD/VCD players does not show any such trend.

All households use MH electricity for lighting and roughly about 10 percent of the households use it in appliances (table 3.3). However, it does not reveal how much of the consumed electricity is used for a specific purpose. For convenience, we divide household's MH electricity use into lighting and non-lighting. As households move from low- to high-income groups, their lighting energy consumption goes up (figure 3.2). But what is more important is how the relative shares of lighting and non-lighting consumption vary as households climb up the income quintiles. Although the level of lighting energy goes up with income growth, the relative share of lighting energy in total energy goes down. Consequently, the share of non-lighting electricity content goes up too—from 2 percent for the lowest quintile households to 4.4 percent for the highest quintile households. This trend is exactly what is expected—first and foremost use of electricity is for lighting but its share goes down as households diversify their electricity consumption.

Figure 3.2: Trend in lighting energy consumption with income change

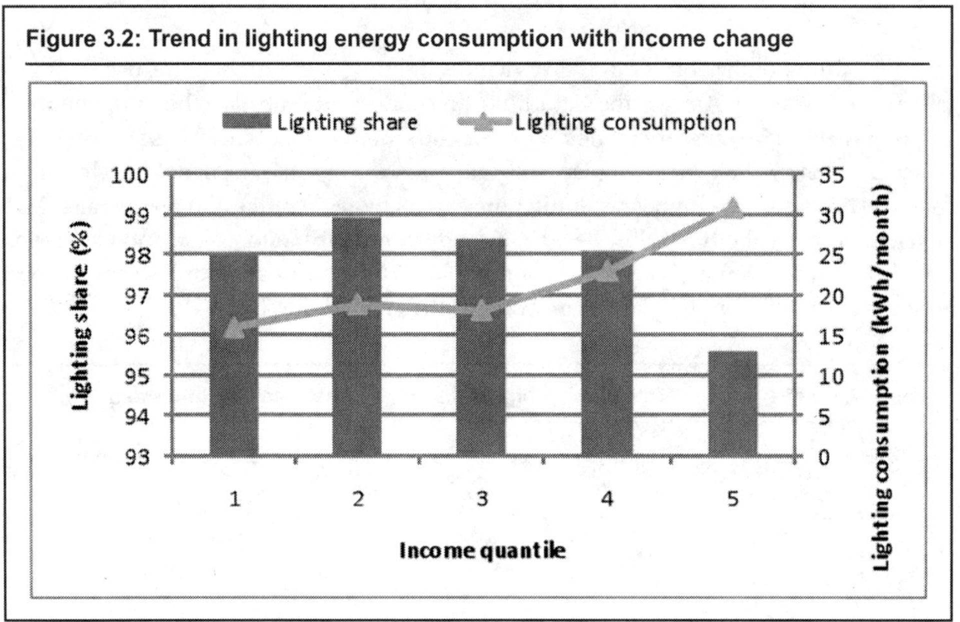

Source: World Bank-AEPC survey 2009.

For the rural poor whose primary concern is to take care of their basic needs, MH electricity use is limited to lighting only (table 3.12). Once they become more solvent they start diversifying their electricity use by acquiring different appliances. However, even with the expected trend, the extent of non-lighting use of electricity by the rural households in Nepal is still very low—only 2.5 percent of the MH electricity consumed is used for non-lighting purposes.

Table 3.12: Distribution of HH MH energy consumption by use type and income quintile

MH use type	1st	2nd	3rd	4th	5th	All households
Lighting (kWh/month)	15.96 (98.0)	18.87 (98.9)	18.10 (98.4)	23.00 (98.1)	30.76 (95.6)	21.49 (97.5)
Non-lighting (kWh/month)	0.33 (2.0)	0.21 (1.1)	0.29 (1.6)	0.44 (1.9)	1.42 (4.4)	0.56 (2.5)
Observations	300	300	300	300	299	1,499

Note: Figures in parentheses are shares of electricity used for lighting and non-lighting.
Source: World Bank-AEPC survey 2009.

What is the level of quality of service among MH households?

For MH households, welfare impacts can only be sustained if the quality of MH service is reasonably good. Service interruptions can be of two types: power outage (scheduled and unscheduled) and voltage fluctuation (table 3.13). Unscheduled power outage is widespread among the MH users in rural Nepal—all households face it on a daily basis. The average duration of such outage is 9 hours per day and it does not vary much by the level of consumption—households that consume very little electricity (less than or equal to 5 kWh/month) are subject to roughly the same level of outage as those who are heavy users (consume over 20 kWh/month). MH households are also subject to unscheduled power outages, mostly because of the monsoon and maintenance. Unlike unscheduled outages, scheduled outages show an increasing pattern as household's energy consumption goes up. There is no scheduled outage for households in the lowest consumption group, while almost 18 percent of the households in the highest group are subject to scheduled outage. The duration of such outage also goes up with electricity consumption, and it is almost 22 days per year for the highest consumption group.

Table 3.13: Power outage and voltage fluctuations in MH service by MH consumption (kWh/month)

Reliability problem type	<=5	5-10	10-15	15-20	20+	All MH users
Unscheduled outage						
HHs affected (%)	100	100	100	100	100	100
Average duration (Hours/day)	8.9	9.3	9.1	8.7	8.8	9.0
Scheduled outage						
HHs affected (%)	0	0.8	2.9	8.1	17.8	7.5
Average duration (Days/year)	-	18.8	19.6	19.5	21.8	21.3
Voltage fluctuations						
HHs affected (%)	53.0	57.4	60.7	53.6	71.6	61.6
Average monetary loss (Rs./month)	2.0	8.2	29.9	18.6	16.7	15.5
Observations	193	373	305	165	463	1,499

Note: The sample is restricted to MH users only.
Source: World Bank-AEPC survey 2009.

Besides power outage, voltage fluctuation also affects a significant number of MH users—almost 62 percent of the MH users. Although the households in the highest consumption group are affected the most (about 72 percent of them), the pattern is not consistent with electricity consumption. Unlike power outage, which simply causes inconvenience or disruptions in household activities, voltage fluctuations can cause real damage to household appliances. The average monetary loss from voltage fluctuation is about 16 Rs. per month, and no clear pattern in monetary loss emerges as the electricity consumption goes up.

Roughly about 72 percent of the MH plant settlements have technicians to address service disruptions, however, the availability of technicians does not make a difference in the incidence of voltage fluctuation (table 3.14). Power outage problem in settlements with technicians is more severe (9.2 hours per day) than that in settlements that do not have technicians (8.3 hours per day). That is somewhat surprising – maybe because the technicians are not trained or skilled enough, or they are not quick enough in addressing a problem.

Table 3.14: Power outage and voltage fluctuations by technician availability in plants

Reliability problem type	Technician available in the MH settlement	Technician not available in the MH settlement	t-stat of the difference
Duration of unscheduled power outage (Hours/day)	9.2	8.3	(7.98)**
Incidence of voltage fluctuation (%)	60.8	63.7	(-1.02)
Observations	1,063	436	1,499

** refers to a significance level of 5 percent or better.
Note: The sample is restricted to MH users only.
Source: World Bank-AEPC survey 2009.

MH households use a variety of alternative arrangements to cope with outages, among which kerosene lamps ranks by far the highest—more than 60 percent of the MH households use kerosene lamps during power outage (table 3.15). It clearly explains why MH households spend a good amount of money (over Rs 92 per month) on kerosene. Among other measures, candle, emergency charger light, and other small miscellaneous measures are also adopted by a good number of households. Only a small share of households (less than 2 percent of the MH users) does not take any measure, that is, they stay in the dark. Table 3.15 shows that these measures do not show any consistent trend as household's electricity consumption goes up. Therefore, the real cost of MH electricity use for rural households in Nepal goes far beyond what the households incur from tariff alone.

Table 3.15: Measures taken by MH users during power outages by MH consumption (kWh/month)

Measures	<=5	5-10	10-15	15-20	20+	All MH users
Use candles (%)	17.1	21.3	13.8	9.9	18.7	17.1
Use kerosene lamps (%)	60.6	46.7	50.7	70.1	78.0	62.4
Use emergency light (%)	9.2	16.2	13.3	29.2	14.6	15.5
Take other measures (%)	15.6	23.4	31.9	17.4	11.0	19.2
Stay in the dark (%)	3.2	2.0	1.6	2.4	1.0	1.8
Observations	193	373	305	165	463	1,499

Note: The sample is restricted to MH users only. Sum of all percentage figures of measures adds to more than 100 as some households take multiple measures.
Source: World Bank-AEPC survey 2009.

What are the factors determining MH connectivity?

Given potential benefits and cost savings of MH, in particular in lighting, it is important to examine what determines MH connectivity at the household level. That is, given that an MH plant is established in a VDC or location, what are the factors that influence certain households to get MH? We look at a number of factors at the household and community levels that determine the MH connectivity. An equation for a household's access to MH can take the following form:

$$E_{ij} = a^e + \beta^e X_{ij} + \gamma^e V_j + \varepsilon^e_{ij} \qquad (1)$$

where E_{ij} is the MH access variable (1 if the household has MH connection and 0 otherwise) for the *i*-th household of the *j*-th community, X_{ij} is a vector of household characteristics (for example, head's age, gender, household landholding, etc.), V_j is a vector of community characteristics (for example, infrastructure and price variables, including the community-level prices of alternate energy sources such as fuel wood, kerosene, etc.), β^e, and γ^e are parameters to be determined, and ε^e_{ij} is the non-systematic error.[5] Since household's access to MH is a binary variable (1 if household has access and 0 otherwise), we use a probit model to estimate it. The results of the estimation and descriptive statistics of explanatory variables are shown in table 3.16.[6]

Education, surprisingly, does not play any role in household's access to MH. Rich households are more able to absorb the cost of connection and tariffs and more likely to connect to MH than poor households—a fact that is reflected from the positive impact of household land on MH access. A 10 percent increase in the household land increases its probability to connect to MH by 0.3 percent. MH connectivity is insensitive to fuel wood price, which is not surprising. As mentioned before, MH is not a substitute for fuel wood. However, kerosene price has a positive impact of MH connectivity, which is also expected. Since both have a primary use for lighting, kerosene and electricity (from MH) are substitutes for each other. An increase of 10 Rs. in kerosene price increases the likelihood of having MH electricity by 1 percent. Among the agroclimate variables, higher altitude lowers adoption of MH, while rainfall increases it.

Table 3.16: Probit estimates of household's access to MH (N=2,497)

Explanatory variables	Means of control variables	Estimates
Sex of household head (M=1, F=0)	0.90 (0.30)	-0.022 (-1.68)*
Age of household head (years)	43.66 (13.25)	0.001 (2.26)**
Education of household head (years)	2.31 (3.53)	0.0004 (0.28)
Highest education among household adult males (years)	4.77 (4.45)	0.001 (0.74)
Highest education among household adult females (years)	2.95 (4.04)	-0.001 (-0.55)
Household size	5.03 (1.82)	0.011 (4.42)**
Log of household landholding (decimals)	1.29 (1.39)	0.030 (2.53)**
Community price of fuel wood (Rs./kg)	3.39 (0.59)	0.013 (0.57)
Community price of kerosene (Rs./liter)	81.03 (10.86)	0.014 (5.18)**
Community price of LPG (Rs./kg)	1,305.42 (23.22)	-0.002 (-3.29)**
Altitude of VDC (100 m)	23.86 (9.57)	-0.008 (-4.05)**
Rainfall in VDC (100 mm/year)	13.89 (5.37)	0.005 (7.84)**
R^2		0.201

Note: Figures in parentheses are standard deviations in Means column and t-statistics in Estimates column.

* and ** refer to a significance level of 10 percent and 5 percent respectively. Explanatory variables additionally include community price of consumer goods, community wage, and infrastructure information.

Source: World Bank-AEPC survey 2009.

What is the MH connectivity among rural enterprises?

When REDP MH initiatives started in rural Nepal during mid-1990s, their focus was domestic customers only. However, it soon expanded to include small micro-enterprises (mainly to support rice milling activities). Since domestic use of MH is mostly for lighting in the evening, production activity can go on during the day without the need for significant capacity increase of the plant. In this section we discuss different features of MH use by small enterprise. Again for the purpose of this study, enterprises have been categorized into two groups by their MH connectivity.

Major activity of the rural enterprises is agro-based food processing. Overall, about 43 percent of enterprises are engaged in such activity, and while 56 percent among MH enterprises are so (table 3.17). About 10 percent are in cottage industry and handicrafts. Trading is also fairly common among the enterprises—32 percent MH and 37 percent non-MH enterprises do trading. A small share of enterprises is engaged in service-oriented activities.

Table 3.17: Distribution of sample enterprise types by MH use (%)

Enterprise activity	MH users	Nonusers of MH	Whole sample
Agro/food processing	56.0	42.2	42.8
Cottage industry and other small manufacturing	9.6	10.0	10.0
Trading	31.7	37.3	37.1
Service industries	2.7	10.5	10.1
Observations	30	35	65

Source: World Bank-AEPC survey 2009.

MH enterprises are relatively new, with an average duration of operation being 6 years, while non-MH enterprises have been in operation for about 9 years (table 3.18). Besides being in business for a shorter duration, MH enterprises are also smaller when measured in terms of physical asset—average asset value of MH enterprises is about Rs. 200,000, while for non-MH enterprises it is more than Rs. 300,000.

Table 3.18: General characteristics of sample enterprises by MH use (%)

Characteristics	MH users	Nonusers of MH	Whole sample
Years of operation	6.4	9.0	8.9
Number of employees	2.3	2.4	2.4
Investment (Rs.)	83,691.3	88,094.5	87,913.4
Land size (acre)	0.017	0.011	0.011
Market value (Rs.)	199,732.3	308,872.6	304,385.8
Observations	30	33	63

Source: World Bank-AEPC survey 2009.

How much is the MH and non-MH enterprise energy consumption?

The consumption of MH electricity and diesel are the two important energy sources that distinguish MH from non-MH enterprises. Diesel provides the production energy for non-MH enterprises—which is why a significant share of non-MH enterprises use diesel, and MH enterprises do not use it at all. A small percentage of MH and non-MH enterprises use biomass. Both use kerosene, but the non-MH enterprises use it more. Finally, both types of enterprises use other sources, namely, solar PV, candle, dry cell, and generator.

Table 3.19: Energy use pattern of sample enterprises (%)

Energy sources	MH users	Nonusers of MH	Whole sample
Biomass	8.5	8.9	8.9
Kerosene	63.1	77.7	77.1
Diesel	0	51.0	48.9
Micro-hydro (MH)	100.0	0	4.1
Other sources	68.8	80.2	79.8
Observations	30	33	63

Note: Biomass includes fuel wood, straw, and rice husk, and other sources include solar PV, generator, candle, and dry cell. Sum of energy share figures adds to more than 100 percent because enterprises often use multiple energy sources.
Source: World Bank-AEPC survey 2009.

Among the different sources that MH enterprises use for their energy needs, most is provided by MH electricity (with a monthly consumption of 21 kgOE), while diesel is the major source for non-MH enterprises (with a monthly consumption of about 34 kgOE) (table 3.20). Among other sources, biomass comes next, providing 19.4 kgOE/month of energy for MH enterprises and 12.15 kgOE/month for non-MH enterprises. Although both types of enterprises use kerosene, its contribution is relatively small in the overall energy demand of the enterprises. Since MH enterprises are smaller than the non-MH ones, it is not surprising that energy consumption of MH enterprises (68.8 kgOE/month) is less than that of their counterpart non-MH enterprises (80.2 kgOE/month).

Table 3.20: Energy consumption of sample enterprises by sources (kgOE/month)

Energy sources	MH users	Nonusers of MH	Whole sample
Biomass	19.41	12.56	12.45
Kerosene	2.37	1.74	1.73
Diesel	0	33.20	32.48
Micro-hydro (MH)	21.40	0	0.88
Other sources	0.19	0.18	0.19
Total	43.37	47.68	47.73
Observations	30	33	63

Note: Biomass includes fuel wood, straw, and rice husk, and other sources include solar PV, generator, candle, and dry cell. Sum of energy share figures adds to more than 100 percent because enterprises often use multiple energy sources.
Source: World Bank-AEPC survey 2009.

MH enterprises spend Rs. 1,037 per month on energy while non-MH enterprises spend Rs. 2,863 per month (table 3.21). Of that, expenditure on MH is Rs. 667 (64 percent of the total energy expenditure) for MH enterprises and expenditure on diesel is 2,492 (87 percent of the total energy expenditure). That is, non-MH households spend proportionately much more on diesel than the energy content of diesel would justify (which is 71 percent of total energy consumption). They also spend much more on miscellaneous small energy sources (about Rs. 130) than their counterpart MH enterprises (Rs. 42), despite consuming about the same amount of energy. Overall, rural enterprises in Nepal spend Rs. 2,786 on energy per month at a price of Rs. 58 per kgOE.

Table 3.21: Energy expenditure of sample enterprises by sources (Rs./month)

Energy sources	MH users	Nonusers of MH	Whole sample
Biomass	162.99	68.29	74.48
Kerosene	164.33	155.20	158.67
Diesel	0	2,492.42	2,396.29
Micro-hydro (MH)	667.42	0	28.31
Other sources	42.34	129.55	128.24
Total	1,037.09	2,863.47	2,786.00
Observations	30	33	63

Note: Biomass includes fuel wood, straw, and rice husk, and other sources include solar PV, generator, candle, and dry cell.
Source: World Bank-AEPC survey 2009.

MH enterprises have an average yearly revenue of more than Rs. 75,000, of which about Rs. 32,000 is profit, while those figures for non-MH enterprises are about Rs. 133,000 and Rs. 40,900 respectively (table 3.22). Since MH enterprises are smaller than non-MH enterprises, it is not surprising that their revenue, business expenditure, and profit are also smaller than that of non-MH enterprises. That is why a direct comparison of these figures between MH and non-MH enterprises may not be meaningful in assessing the role of MH. In contrast, we create two more indicators, namely, profit as a percentage of the revenue and profit as a percentage of the enterprise asset, which are more comparable between the two types of enterprises than the absolute figures of revenue or profit. The profit of MH enterprises is 39.5 percent of their revenue, while that for non-MH enterprises is 37.8 percent of their revenue. Similarly, the profit is 19.5 percent of the enterprise value for MH enterprises and 21.5 percent for non-MH enterprises. The t-statistics of the differences of these two figures is insignificant, implying no difference between MH and non-MH enterprises. It seems that the adoption of MH by the enterprises has not made a difference yet in profit-making, possibly because they are smaller, have been in operation for a shorter period, and consume less energy than their counterpart non-MH enterprises.

Table 3.22: Financial characteristics of sample enterprises by MH use

Characteristics	MH users	Nonusers of MH	Difference	t-stat of difference
Revenue (Rs./year)	75,298.6	132,636.6	-57,338.0	(-1.04)
Business expense (Rs./year)	42,931.4	91,758.3	-48,826.9	(-0.98)
Profit (Rs./year)	32,367.2	40,878.3	-8,511.1	(-0.54)
Profit as a percentage of revenue (%)	39.5	37.8	1.76	(0.14)
Profit as a percentage of enterprise value (%)	19.5	21.5	-2.0	(-0.14)
Observations	30	33	63	

Source: World Bank-AEPC survey 2009.

The electricity use pattern of MH enterprises can be examined by dividing them into 5 quintiles based on their value (table 3.23). As enterprises get bigger, their consumption of electricity and expenditure on MH electricity go up monotonically, so does their peak load. While the lowest quintile MH enterprises consume 74 kWh/month, those in the highest quintile consume more than 404 kWh/month. However, their peak load goes up much faster than their consumption—the peak load of the lowest quintile enterprises is about 180w, while that of the highest quintile enterprises is about 6,075w. As a result, the load factor of the enterprises goes down as they grow bigger (from 0.57 to 0.09), indicating that the electricity consumption of MH enterprises becomes less optimal as they grow bigger.[7]

Table 3.23: Electricity use pattern by enterprise value quintile for MH enterprises

Electricity use	1st	2nd	3rd	4th	5th
Consumption (kWh/month)	73.5	193.3	232.2	397.5	404.3
Expenditure (Rs./month)	107.2	682.0	770.9	837.8	1,088.2
Peak load (w)	180.8	1,278.9	2,908.1	3,475.3	6,075.6
Load factor	0.565	0.210	0.111	0.159	0.092

Source: World Bank-AEPC survey 2009.

As the enterprises grow bigger, their revenue and expenditure tend to go up, although not consistently (table 3.24). Enterprise profits follow the same pattern. Lowest quintile enterprises have an average revenue of Rs. 38,149 and a profit of Rs. 11,426, while for the highest quintile enterprises those figures are Rs. 122,012 and Rs. 49,221, respectively. Share of profit as a percentage of revenue does not follow a consistent pattern either—it is 29 percent for lowest quintile enterprises and 38 percent for highest quintile enterprises. Finally, share of profit as a percentage of enterprise seems to decrease as enterprise asset value goes up, probably implying a diminishing return.

Table 3.24: Financial outcomes by enterprise value quintile for MH enterprises

Electricity use	1st	2nd	3rd	4th	5th
Revenue (Rs./year)	38,148.7	91,315.9	81,739.1	62,339.9	122,012.1
Business expense (Rs./year)	26,723.2	42,673.0	46,863.9	35,566.6	72,790.8
Profit (Rs./year)	11,425.5	48,642.9	34,875.2	26,773.3	49,221.3
Profit as a percentage of revenue (%)	0.29	0.53	0.42	0.39	0.38
Profit as a percentage of enterprise value (%)	0.20	0.36	0.20	0.11	0.12

Source: World Bank-AEPC survey 2009.

How do MH enterprises cope with power outages?

The quality of MH service for the enterprises is not very satisfactory (table 3.25). The incidence of power outage is 100 percent for the rural MH enterprises in Nepal. The average duration of power outage is 8 hours and it does not vary much across enterprise quintiles. Besides power outage, enterprises suffer from voltage fluctuations as well. More than half of the enterprises report that they face voltage fluctuation problems. There is no consistent pattern of voltage fluctuations with enterprise value quintiles.

Table 3.25: MH service quality by enterprise value quintile for MH enterprises

Measures	1st	2nd	3rd	4th	5th	All enterprises
Incidence of power outage (%)	100.0	100.0	100.0	100.0	100.0	100.0
Average duration of outage (hours)	8.8	9.0	7.6	7.5	7.1	8.0
Incidence of voltage problem (%)	30.3	93.1	20.4	88.7	43.1	55.2

Source: World Bank-AEPC survey 2009.

A majority of enterprises use kerosene lamps, followed by those who use emergency light during power outages (table 3.26). A good number of MH enterprises use candles, and some stay in the dark. There is no clear pattern in the measures adopted across the quintile of enterprise asset value. Measures taken by the enterprises can satisfy their lighting needs only—that is, no measures are taken to compensate for the loss of processing energy during a power outage. For example, there is no mention of generator use, which could possibly power the tools or machinery supported by MH electricity. This clearly shows that MH enterprises, for one reason or another, do not invest in alternate electricity sources despite the ubiquitous nature of power outages. This may be another reason that having MH does not improve the enterprise's bottom line (that is, profit), as non-MH enterprises do not have to cope with any significant interruptions in the energy flow—the energy they consume is of inferior quality no doubt, but at least they have it when they need it.

Table 3.26: Measures taken by MH enterprises during power outage

Measures	1st	2nd	3rd	4th	5th	All enterprises
Use candles (%)	9.0	45.1	62.8	17.7	17.0	29.6
Use kerosene lamps (%)	91.0	55.9	41.1	70.9	56.4	64.6
Use emergency light (%)	69.7	62.8	36.8	46.8	0	45.6
Stay in the dark (%)	0	14.7	32.1	11.3	17.0	14.4

Note: Sum of the percentage figures adds to more than 100 because households adopt more than one measure.
Source: World Bank-AEPC survey 2009.

Notes

1. Since MH households and enterprises have been purposely overdrawn in the sample, all descriptive and econometric analyses of this study have been weighted to make them representative of the rural areas of Nepal.
2. However, because of this limitation in capacity, lighting needs of some households may not be satisfied and they resort to backup lighting such as kerosene, candle, etc. Households also use backup lighting during the disruptions of MH service. We cover service disruptions later when we discuss MH use in detail.
3. It must be noted that here only 12 percent of the biomass expenditure is for actual purchase (that is, households paid for), and the rest is calculated from (attributed to) produced and collected quantities using market price of biomass.
4. Twenty-two percent is relatively high for energy expenditure-income ratio. However, this is not impossible, because a large share of energy expenditure is for biomass, and as mentioned before, a bulk of the biomass expenditure is inputted (that is, not actually paid for by the households) as it is mostly collected or produced. Since poor households collect a larger share of their biomass than the

wealthy households, the unspent component in the energy expenditure is higher for poor households than that of their counterpart wealthy households. For example, unspent component that contributes to total biomass energy expenditure is almost 92 percent for the lowest quintile households and 77 percent for the highest quintile households. It is this unspent component that makes the energy expenditure-income ratio somewhat unrealistically high for the poor households.
5. Household landholding, instead of income, is included as a control variable, because, household income may be influenced by MH connectivity. And unlike income, landholding does not change in the short term and therefore is not likely to be changed due to the access to MH.
6. Results show the impacts of selected control variables only. Not shown in Table 4.18 but included in the regression are community prices of consumer goods and wage variables.
7. Load factor is the ratio of average load to peak load in a period. A high load factor means more optimal use of electricity and greater output.

CHAPTER 4

Benefits of Electrification to Rural Households

This chapter presents the results of the evaluation of benefits of micro-hydro electrification in rural Nepal across multiple dimensions using the World Bank-AEPC Survey, 2009, carried out as part of this study. We begin by estimating the consumer surplus, which can be represented as the lower bound of benefits. We then move on to evaluate the MH benefits for multi-dimensional welfare outcomes such as health, education, fertility, women's empowerment, and greenhouse gas (GHG) emissions. The propensity score matching (PSM) technique has been employed to determine the statistical significance and robustness of such benefits (the technique is described in Annex 4).

How do the households benefit from MH connectivity?

Benefits of electricity are well recognized in the development community (World Bank 2008). Access to modern energy, in particular electricity, is key to the fulfillment of several time-bound targets outlined in the Millennium Development Goals (MDGs) set in 2000 at the United Nations Millennium Summit (United Nations 2005). Electrification benefits at the household level start accumulating immediately with lighting, which is the primary use for most rural households. The second common use is TV watching, and together, lighting and TV watching constitute 80 percent of rural electricity consumption (IEG 2008).

Electricity provides light that is hundreds of times brighter and at the same time cheaper than kerosene-based lighting. Lighting with electricity allows business activities to extend well beyond daylight hours, which has the potential for employment and income growth. It enables women to be engaged in productive activities in the evening, and increases their exposure to the outside world and education through electronic media such as TV powered by electricity. Electric lighting may also increase study hours for school-going children, increasing their educational achievements (Barnes, Peskin and Fitzgerald 2003; Kulkarni and Barnes 2004).

Measuring the benefits of electrification has undergone an evolution in the past 40 years. While it was always claimed that benefits outweigh the costs, the evidence was not robust as only the limited impacts were considered. Overtime, the approach has expanded from consumer surplus that serves as a lower bound of benefits to broad outcomes such as health, education, women's empowerment, climate change etc. that robustly quantify the benefit flows and ensure the cost-effectiveness of the investments (Barnes 2008) (figure 4.1).

Figure 4.1: Evolution in rural electrification benefits evaluation

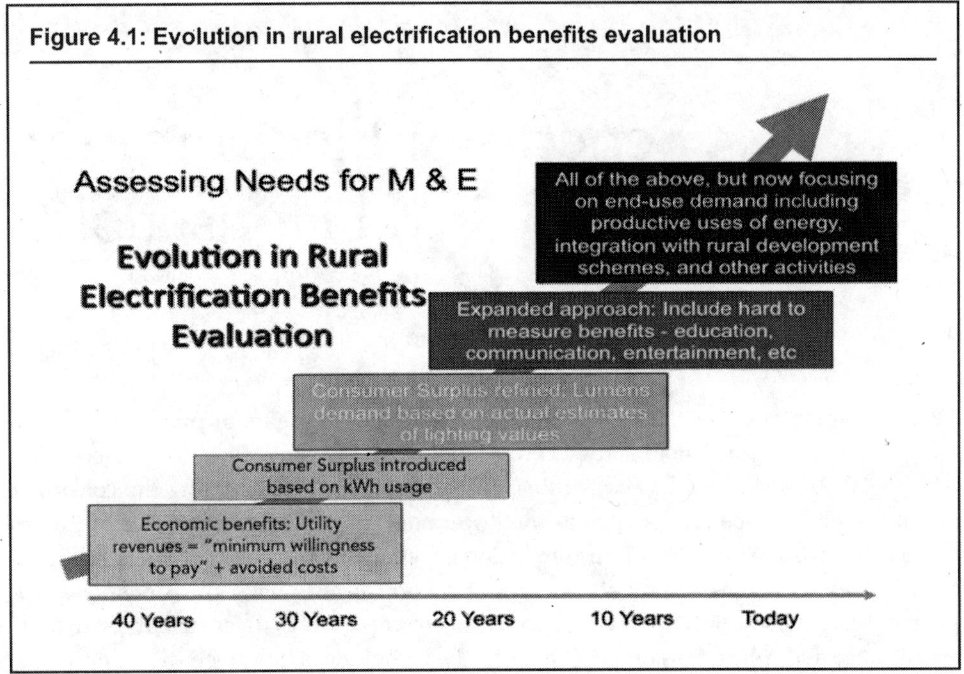

Source: Barnes (2008): Presentation "Monitoring and evaluating impacts of rural electrification: Past experience and way forward." Energy Week. The World Bank.

Consumer surplus

The primary benefit of MH is lighting as households benefit from better quality of lighting (table 4.1). For an average household the quantity of electric light from MH is more than 100 times brighter than the light available from kerosene lamps. The amount of light (in kilo-lumen-hr/month) used by MH households increases from low- to high-income households, and for the highest income quintile the amount of light from MH electricity is more than two-hundred times that from kerosene used by non-MH households.

Table 4.1: Lighting intensity from kerosene and MH electric lighting by income quintile (klumen-hr/month)

Lighting source	1st	2nd	3rd	4th	5th	All households
Kerosene lamps (non-MH HHs)	2.0	2.4	2.6	2.1	2.1	2.2
Electricity (MH HHs)	203.3	266.3	214.2	297.7	410.5	287.8
Observations	500	500	499	500	498	2,497

Source: World Bank-AEPC survey 2009.

Households either replace or reduce kerosene consumption after getting MH connection (figure 4.1). This replacement or reduction of kerosene saves households money even when the cost of MH electricity is taken into account. MH households spend in general less on kerosene and MH combined than what non-MH households spend on kerosene. Overall, non-MH households spend about Rs. 200 on kerosene, while MH households spend Rs. 154 on MH and kerosene combined (figure 4.2). Replacing kerosene with electricity is not just cost-saving, the quality of light achieved from electricity is much better than that possible from kerosene.

Figure 4.2: Comparison of kerosene and electricity expenditure between MH and non-MH HHs by income quintile

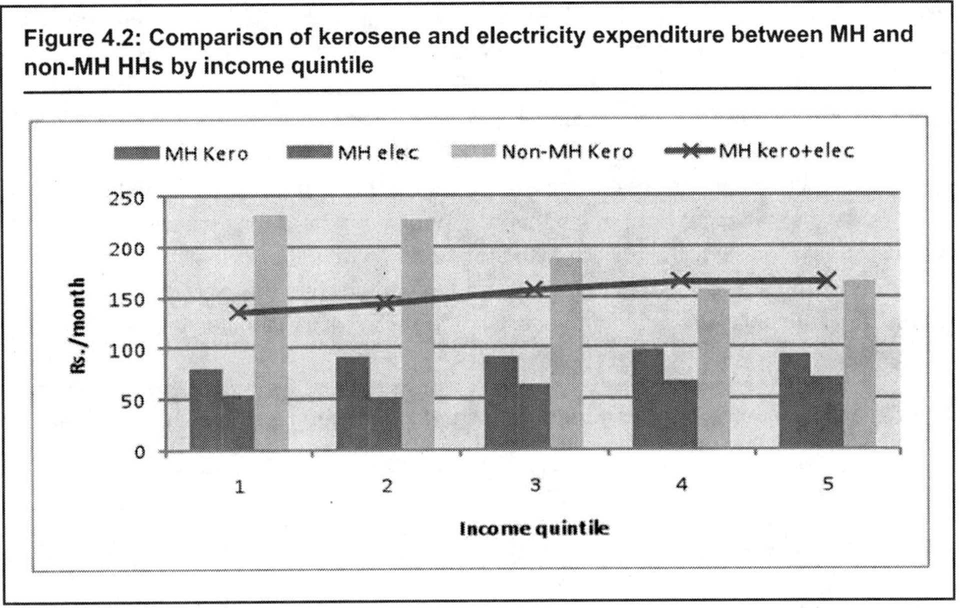

Source: World Bank-AEPC survey 2009.

The consumer surplus measures the potential savings from the quality of lighting that MH households accumulate. Consumer surplus is a conventional method of determining intervention benefits accruing from the consumption of goods and services. It is defined as the difference between the total amount the consumers are willing to pay (WTP) for a good or service and the total amount they actually pay (AP), provided WTP is higher than AP. Consumer surplus is the (virtual) savings accumulated by the consumers from not paying the extra amount they are willing to and can pay.

When households switch to MH they enjoy consumer surplus for different services, most important among them is lighting. Households enjoy consumer surplus by switching to MH because they receive lighting service at a cheaper price (AP) than what they would have paid for using kerosene lighting (WTP). The reason that AP is lower than WTP is the fact that unit price for lighting (in Rs./klumen-hr) is much cheaper for MH than it is for kerosene. Two assumptions need to be made in the measurement of consumer surplus for switching from kerosene to MH. First, determining WTP for MH users is tricky as we do not have that information readily available, nor do we have the information on the amount that current MH users used to pay for kerosene lighting before switching to MH. In that case, the amount that current non-MH users pay for kerosene lighting is used as a proxy for all practical purposes. Second, the calculation of exactly how much current MH users and nonusers pay for lighting is not straightforward, because households use kerosene or MH also for non-lighting purpose (although in small amount) and it is difficult to disaggregate their spending by lighting and non-lighting uses. This is further complicated by the fact that households often use other sources for lighting too in addition to kerosene or MH (for example, candles). To disentangle these linkages, we consider only those non-MH users who use kerosene alone for lighting service and those MH users who use MH electricity alone for lighting service, and none of them use kerosene or MH for any other purposes.

Kerosene is 400 times more expensive than MH when measured by lighting intensity. The cheaper price enables MH households to enjoy more light and generous consumer surplus than their counterpart non-MH households. MH households in rural Nepal enjoy a consumer surplus of almost Rs. 700 (table 4.2). The trend in consumer surplus by household income is not consistent. That may be because households do not use MH as much as they can for the price they pay. Households are charged by the number of light bulbs they use regardless of the duration of lighting use. If rich households have more (or higher wattage) light bulbs than the poor households, they have the potential of consuming more lighting service, which may result in higher consumer surplus. The consumer surplus calculated here may be underestimated—a more accurate estimation of consumer surplus is possible if the usage of MH electricity is metered.

Table 4.2: Consumer surplus for lighting service enjoyed by HHs switching from kerosene to MH by income quintile

Variables	1st	2nd	3rd	4th	5th	All households
Price paid for kerosene lighting (Rs./klumen-hr)	92.2	81.2	76.2	88.3	78.3	84.1
Quantity of kerosene lighting consumed (klumen-hr/month)	2.94	2.06	2.25	2.21	2.54	2.38
Price paid MH lighting (Rs./klumen-hr)	0.18	0.16	0.28	0.15	0.29	0.21
Quantity of MH lighting consumed (klumen-hr/month)	283.05	342.63	185.53	377.50	282.38	284.42
Consumer surplus (Rs./month)	826.0	629.7	562.0	712.0	739.7	697.2

Source: World Bank-AEPC survey 2009.

Consumer surplus as a percentage of income decreases consistently from low- to high-income quintile households (figure 4.3). The poorest quintile saves almost 50 percent of their income. There is a trend of diminishing return as a household's income goes up. However, it must be noted here that consumer surplus is not a household's actual savings, it is rather potential savings for not paying the maximum amount it can pay.

Figure 4.3: Share of consumer surplus in income by income quintile

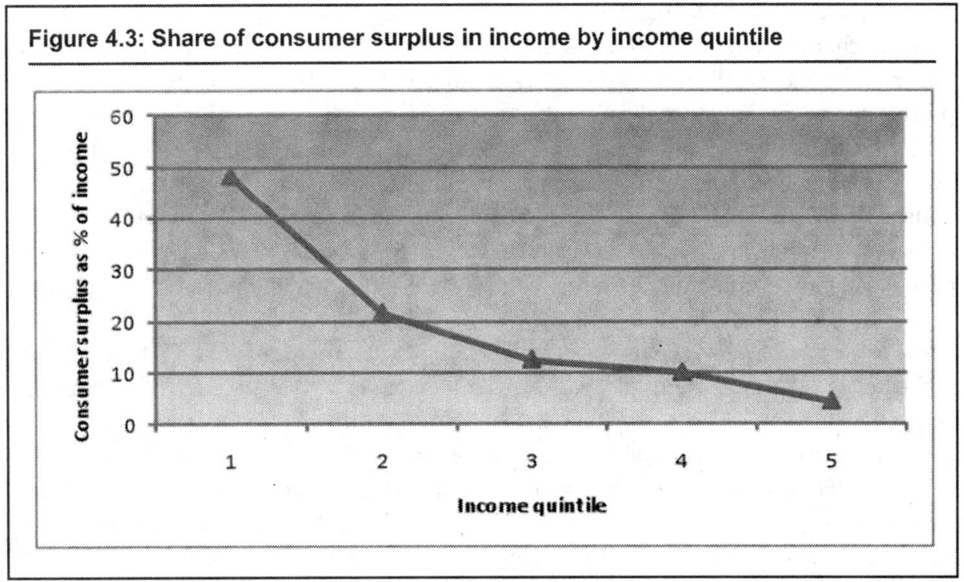

Source: World Bank-AEPC survey 2009.

Econometric estimates

Consumer surplus is only the lower bound of benefits—the scale of impacts is wide-ranging. Using the household survey, we compare welfare outcomes of the MH households with that of control households, and the difference in the means of outcomes between the two types of households provides an estimate of the electrification benefits accrued by MH households. However, benefit estimated this way may be biased, because MH and non-MH households may vary fundamentally in other characteristics, not just in MH connectivity, and it is difficult to determine whether the estimated benefits are due to MH access or other characteristics. This problem can be addressed by using a matching technique, in particular, propensity score matching (PSM) technique, which is the most common among the matching techniques.

Using PSM technique, MH households are matched against the control households so that they are similar in observable characteristics except for their electrification status. PSM technique calculates, for both treated (MH) and untreated (non-MH) households, the probability of treatment or MH-connectivity (called propensity score) as a function of those characteristics from a logit or probit model. Based on the propensity score, treated and untreated households are matched and those not matched are discarded. Finally, the difference in outcomes between the matched treated and untreated households gives estimate of the MH impact. A technical introduction of the PSM technique is given in Annex 4. The World Bank-AEPC survey provides a wide range of household and community characteristics that have been used in the probit regression of the PSM estimation to facilitate the computation of the propensity score. To be more specific, following household variables have been used: head's age, education, gender, household size, maximum education of adult males and females in the household, and household's land asset. At the community level, prices of alternate fuels (fuel wood, kerosene, LPG), daily labor market wages of males and females, prices of different consumer goods (rice, wheat, lentil, meat, fish, vegetables, and so on) and infrastructure information (presence of schools, markets, paved roads, health center) have been used. Finally, two agroclimate variables (altitude of the VDC and yearly rainfall) have also been used in the probit estimation. Together these control variables are expected to provide a reasonably robust estimate of MH connectivity. From the PSM estimation, 429 households from the non-MH samples did not satisfy the matching criteria and so are dropped, leaving 2,070 households (1,499 MH and 571 non-MH) for final comparison. Since these households are exactly similar in their observable characteristics except for MH connectivity, the difference in their outcomes is assumed to have been contributed by MH connectivity.

Two observations need to be made before MH benefits on household outcomes are reported. First, PSM has its limitations. Since, it is based on the observed characteristics of the households it may provide biased estimates of MH benefits if the unobserved characteristics vary for MH and non-MH households and influence the outcomes. Bias arises because we cannot measure and separate the effects of those unobserved characteristics from the MH impacts estimated by PSM. One way to control for such unobserved biases is to use a two-stage instrumental variable technique, which is dependent on finding the right instrument (s).[1] However, we cannot use instrumental variable regression to estimate MH impacts because of lack of instruments. We assume that after controlling for a wide range of observable characteristics in a reasonably large sample, the bias due to unobservable characteristics would not be significant. Second, MH intervention in rural Nepal is accompanied by other inputs as part of REDP's community mobilization effort, for example, capital mobilization, skill enhancement, technology promotion, vulnerable community empowerment etc. Apart from the access to electricity, these inputs are also likely to influence household outcomes. As a result, the estimated impacts of MH in this study are not limited to only electricity.

Economic outcomes

Economic returns on MH connectivity can come in multiple ways. Because of the lighting provided by MH, households can continue their domestic income generation activities well into the evening, women can do sewing or other small-scale activities after daytime household chores, and electronic media like TV and radio can bring them exposure to new entrepreneurial skills, and so on. We focus on four economic outcomes: farm, nonfarm, and total income, and expenditure, all per capita measures. The differences are statistically significant only for two outcomes—a household's nonfarm income and per capita expenditure.

These two outcomes also show positive and significant MH impacts according to PSM estimates (table 4.3). MH connectivity increased household's nonfarm income by 11 percent and expenditure by 9 percent. The gain in nonfarm income, however, is not enough to result in a statistically significant gain in total income. As the consumption level of MH electricity is low and mostly used for lighting, it may take a while for MH to have a robust impact on economic outcomes.

Table 4.3: Impacts of MH access on economic outcomes (Rs./capita/month)

Outcomes	MH HHs	Non-MH HHs	Difference (%)	PSM estimates (nearest neighbor match)
Total income	1,894.4	1,561.4	0.109 (1.56)	-0.048 (-0.53)
Farm income	1,029.2	932.2	0.153 (1.43)	0.183 (1.31)
Nonfarm income	865.2	629.2	0.353 (2.53)**	0.112 (1.91)*
Expenditure	1,456.2	1,263.1	0.039 (1.92)*	0.090 (3.26)**

Note: Figures in parentheses are t-statistics, and * and ** refer to significance levels of 10 and 5 percent respectively.
Source: World Bank-AEPC survey 2009.

Educational outcomes

MH impacts on educational outcomes can probably accrue faster than that on other outcomes. Lighting provided by MH electricity allows school-going children to study more comfortably and conveniently in the evening than what is possible with kerosene-based lighting. Students study for longer hours and have the potential to perform better than children in non-MH households. To underscore the change in study time, figures 4.4.A and 4.4.B show the evening study time of boys and girls by grade for MH and non-MH households. There is no systematic trend in the relationship between grade level and study duration. However, girls from MH households are found to study more in the evening than those from non-MH households for almost all grades. This pattern holds for boys in lower grades.

Figure 4.4.A: Boys' evening study minutes by grade for MH and non-MH HHs

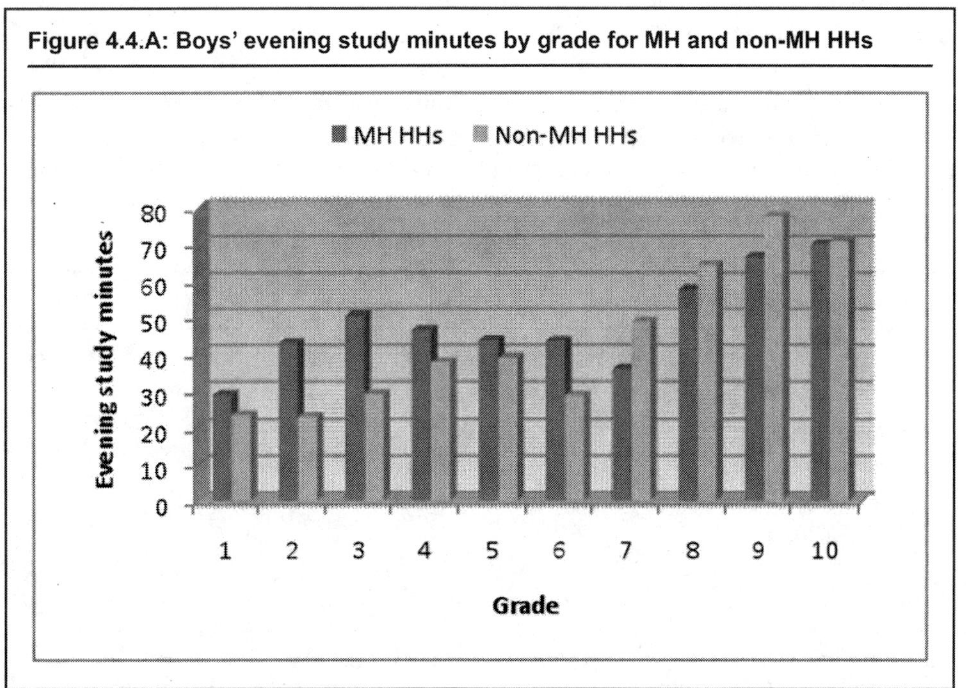

Source: World Bank-AEPC survey 2009.

Figure 4.4.B: Girls' evening study minutes by grade for MH and non-MH HHs

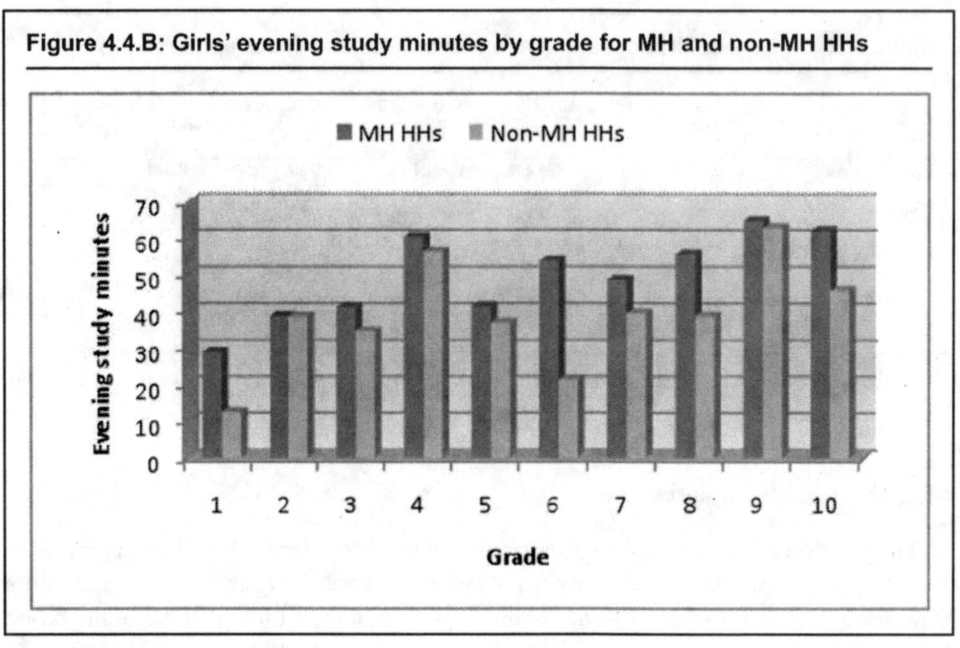

Source: World Bank-AEPC survey 2009.

Boys' educational outcomes are slightly better than that of the girls regardless of MH connectivity, and children from MH households do better than those from non-MH households (table 4.4). Boys from MH households have completed 4.5 years of schooling, while those from non-MH households have completed 4.1 years of schooling. However, their difference is not statistically significant, nor is the PSM estimate of impact. In girls' schooling, on the other hand, a significant difference is observed between MH and non-MH households. Based on PSM estimate, girls from MH households have completed about 0.24 year more of schooling than their counterpart girls from the non-MH households. As for study time in the evening, both boys and girls from MH households do better than their counterparts from non-MH households. For example, evening study time for boys is 50 minutes for MH households and 34 minutes for non-MH households. PSM estimate of their difference is almost 8 minutes, which is statistically significant. So we can conclude that MH connectivity improves children's education.

Table 4.4: Impacts of MH access on educational outcomes (age 5–18)

Outcomes	MH HHs	Non-MH HHs	Difference	PSM estimates (nearest neighbor matched)
<u>Schooling years completed</u>				
Boys	4.53	4.12	0.405 (1.50)	-0.088 (-0.47)
Girls	4.28	3.73	0.551 (1.97)*	0.240 (1.65)*
<u>Evening study (minutes/day)</u>				
Boys	50.1	33.9	16.2 (3.63)**	7.7 (2.32)**
Girls	39.7	30.0	9.7 (2.19)**	12.0 (5.06)**

Note: Figures in parentheses are t-statistics, and * and ** refer to significance levels of 10 and 5 percent respectively.

Source: World Bank-AEPC survey 2009.

Health outcomes

Respiratory and gastronomical are the two major health outcomes affected by MH connectivity. An individual is considered affected by these health problems if she or he is not able to carry out productive work, schooling, or domestic chores. There is not much difference in the suffering of adult males or females—among the households with MH and those without MH, about 5 percent have adult male members who suffer from at least one of the two health problems. Women's suffering follows about the same pattern. However, the difference is the starkest among boys and girls. The share of non-MH households with boys suffering is about 6 percent, while that figure for MH households is about 2 percent.

We present an age-specific breakdown among males and females to examine more closely if the young population (age less than 18) suffer more than the adults (age 18 or above) separately from the two health problems (table 4.5). An adult male from an MH household is affected for 6.2 hours a month, while one from a non-MH household suffers for 5.4 hours a month. However, their difference is not significant. On the other hand, an adult woman from a non-MH household suffers almost double (9.7 hours per month) the hours that her counterpart from an MH household suffers (5.1 hours per month). Again, their difference is not statistically significant. However, PSM estimates show a negative impact—MH connectivity lowers the average monthly suffering of adult women from respiratory diseases by more than 3 hours. Women stay home longer than men and are more likely to be exposed to indoor pollution if kerosene is burnt, either in lamps or in stoves. In non-MH households, this is a distinct possibility where kerosene is probably the only lighting energy. PSM estimates of impacts for boys and girls are also negative, and higher for girls. For example, MH access lowers girls' average suffering from respiratory diseases by 6 hours per month, and boys' suffering by 1.6 hours per month.

Figure 4.5: Share of HHs with members suffering from respiratory or gastrointestinal problems by MH connectivity

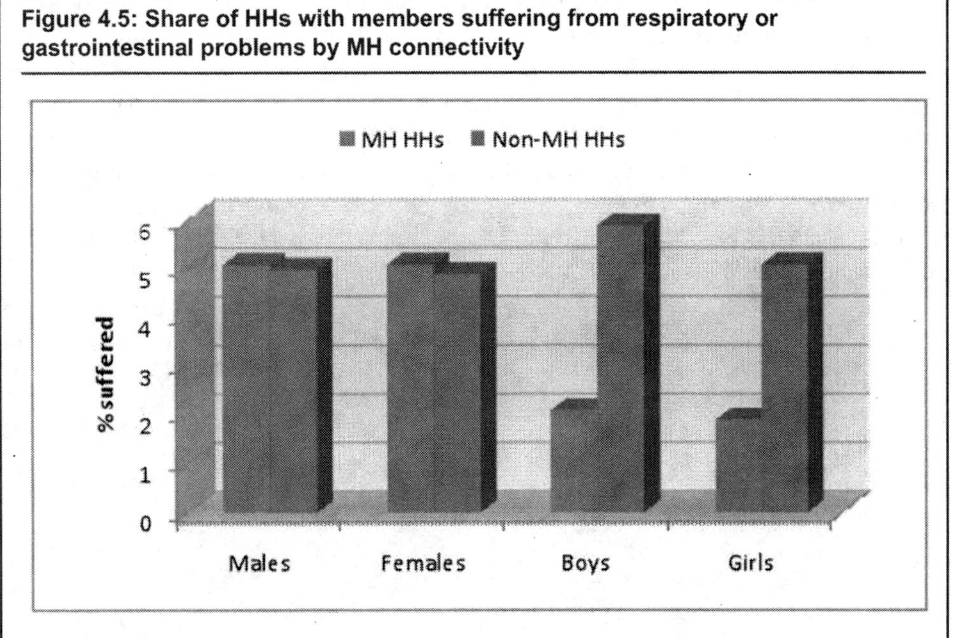

Source: World Bank-AEPC survey 2009.

Adult males from MH households suffer less (about 1.2 hours less per month) from gastrointestinal diseases than those from non-MH households but the difference is not significant (table 4.5). PSM estimates show significant impact only for girls (1.43 hours per month).

Table 4.5: Impacts of MH access on HH health outcomes (hours per month suffered)

Outcomes	MH HHs	Non-MH HHs	Difference	PSM estimates (nearest neighbor matched)
Respiratory problems				
Men (age=>18)	6.2	5.4	0.84 (0.29)	3.7 (0.14)
Women (age=>18)	5.1	9.7	-4.62 (-0.91)	-3.4 (-3.22)*
Boys (age<18)	1.4	5.1	-3.63 (-1.75)*	-1.6 (-2.28)**
Girls (age<18)	1.3	8.2	-6.90 (-1.93)*	-6.1 (-2.82)**
Gastrointestinal (GI) problems				
Men (age=>18)	0.9	2.1	-1.18 (-0.70)	-0.21 (-0.29)
Women (age=>18)	2.2	4.7	-2.41 (-0.80)	-1.7 (-1.03)
Boys (age<18)	1.0	0.9	0.10 (0.08)	-0.28 (-0.34)
Girls (age<18)	0.3	1.7	-1.40 (-0.87)	-1.43 (-1.71)*

Note: Figures in parentheses are t-statistics, and * and ** refer to significance levels of 10 and 5 percent respectively.
Source: World Bank-AEPC survey 2009.

Exposure to media is one of the ways MH connectivity can contribute to better health outcomes. TV, for instance, increases awareness of health issues, in particular, measures that household members can take to prevent or cure health problems. After establishing the evidence that MH connectivity improves health outcomes of the household members, we would like to investigate if having a TV in the household makes a further difference. TV is important when it comes to health outcomes. Incidence of respiratory and gastrointestinal problems among women, boys and girls in MH households with a TV is much less than that in MH households that do not own a TV (figure 4.6). However, men in MH households with a TV seem to suffer more than their counterpart men from MH households without a TV. It may be because men often remain outside the home and watch less TV than women. As expected, women's exposure to a TV contributes to children's health outcomes too.

Figure 4.6: Trend in respiratory and gastrointestinal problems in MH households with TV ownership

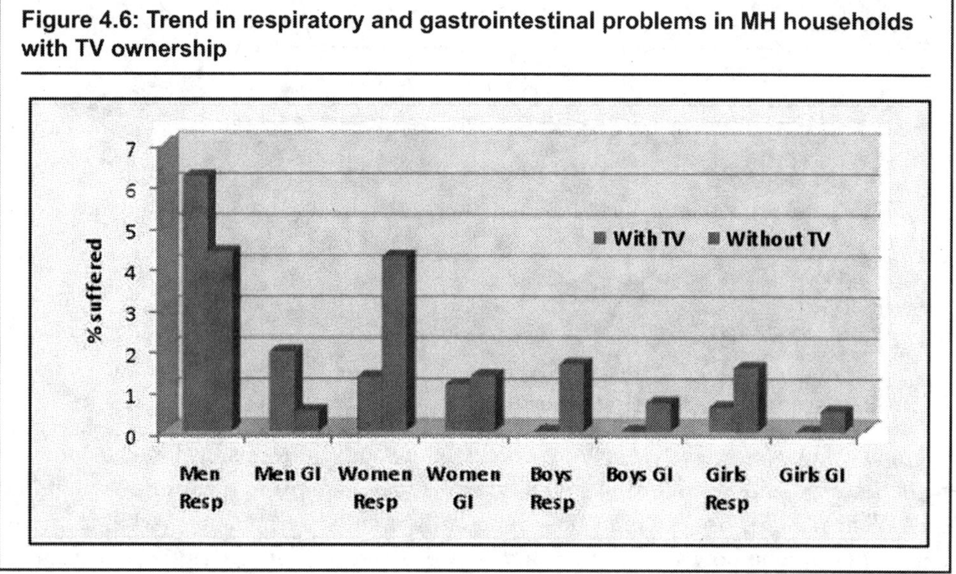

Source: World Bank-AEPC survey 2009.

Women's fertility outcomes

Fertility outcomes are mixed—contraceptive use is slightly better and recent fertility is higher for MH households (table 4.6). Two of the most commonly used measures of women's fertility outcomes are contraception prevalence rate, which is defined as the share of married or sexually active women, usually between ages of 15 and 49, who use any sort contraceptive, and recent fertility, which is defined as the share of married or sexually active women, usually between the ages of 15 and 49, who gave birth during the preceding three years. For example, about 74 percent of the women from MH households use some sort of contraception method, as opposed to 72 percent of women from non-MH households. Furthermore, 26 percent of the women from MH households gave birth during the preceding three years, while such women constitute 23 percent among non-MH households. The difference in either outcome between MH and non-MH households is not statistically significant. PSM estimates, however, show positive impacts in the contraceptive prevalence rate—MH access increases the contraceptive prevalence rate by 3.8 percentage points. This higher prevalence rate has not yet resulted in lower recent fertility—it is possible that these women, who reportedly use contraceptives, do not use them long enough to have an impact on fertility.

Table 4.6: Impacts of MH access on women's fertility outcomes (ages 15–49)

Outcomes	MH HHs	Non-MH HHs	Difference	PSM estimates (nearest neighbor matched)
Contraceptive prevalence rate	0.744	0.718	0.026 (0.59)	0.038 (2.78) *
Recent fertility	0.262	0.233	0.029 (0.70)	0.038 (1.31)

Note: Figures in parentheses are t-statistics, and * and ** refer to significance levels of 10 and 5 percent respectively

Source: World Bank-AEPC survey 2009.

Women's empowerment outcomes

Women's empowerment has been defined in this study in terms of three attributes—their time use, independence in mobility, and independence in decision-making (box 4.1). Women's time use is a critical element of their empowerment, and what is important is to examine how the relative shares of different activities change as a result of an intervention such as MH. Women's time is used in five major groups of activities: household chores (including nurturing of children), income generation activities (IGA) (self- or family-owned), study, leisure, and rest. Household chores take up the highest share of women's time regardless of MH-connectivity (38 percent). Apparently, for any activity, there is not much difference in the activity share by MH access. However, when we look at the actual duration of the activity, such difference becomes statistically significant. Women from MH households spend more time in income generation, study and leisure activities than their counterpart women from non-MH households, and their differences are statistically significant (table 4.7). PSM estimates also show that MH access has enabled women to use their time more productively in IGA and study, while allowing them more leisure time.

Box 4.1: Indicators of women's empowerment

Empowerment type	Subcategories	Definition
Time use	IGA	Time used in income generation activities and marketing
	Study	Time spent in reading and studying
	Leisure	Time spent in watching TV, listening to radio, reading for entertainment, and socialization.
	Rest	Time spent in resting, relaxing, napping, and sleeping
Mobility	Mobility type 1	Independence in visiting often or occasionally places of friends, neighbors, and relatives
	Mobility type 2	Independence in visiting often or occasionally markets, health centers, community meetings, and other places
Decision-making	Fertility issues	Wife is involved in decision-making (either independently or with husband) in issues related to contraceptive use and having child
	Children's issues	Wife is involved in decision-making (either independently or with husband) in issues related to children's education, health, and marriage
	Monetary issues	Wife is involved in decision-making (either independently or with husband) in issues related to spending money

Source: Authors' elaboration.

Figure 4.7: Distribution of women's activities by MH connectivity (%)

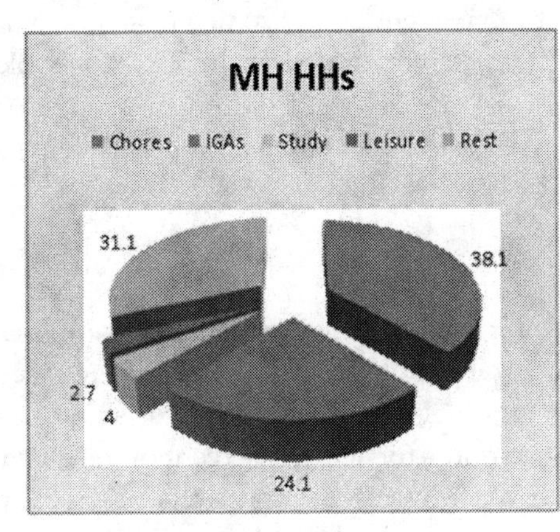

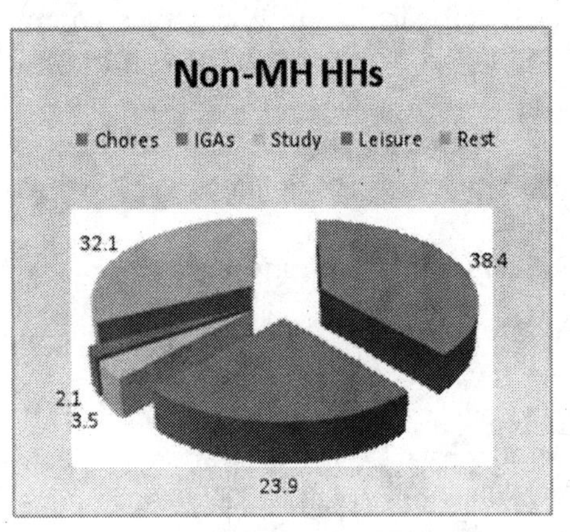

Source: World Bank-AEPC survey 2009.

Type 1 mobility of women from MH households is higher than that of women from non-MH households. More precisely, the incidence of visiting friends, relatives, and neighbors is almost 22 percentage points higher for women from MH households than that for women from non-MH households. However, for mobility type 2 the difference between MH and non-MH households is not statistically significant. Also according to PSM estimates, MH does not have any impacts on women's mobility.

Table 4.7: Impacts of MH access on women's empowerment outcomes

Outcomes	MH HHs	Non-MH HHs	Difference	PSM estimates (nearest neighbor matched)
Time use (hours/day) in Income generating activities	5.81	5.54	0.27 (1.99)**	0.19 (1.97)*
Study	0.96	0.79	0.17 (1.13)	0.20 (1.86) *
Leisure activities	0.71	0.48	0.23 (4.25)**	0.21 (5.71)**
Rest	7.36	7.36	-0.004 (-0.03)	-0.06 (-0.76)
Independence in mobility Have mobility type 1	0.569	0.354	0.215 (4.12)**	-0.013 (-0.36)
Have mobility type 2	0.562	0.504	0.058 (1.07)	0.014 (0.41)
Independence in decision-making Decision-making in fertility issues	0.844	0.726	0.117 (2.44)**	0.042 (1.85)*
Decision-making in children's issues	0.942	0.921	0.021 (0.70)	0.027 (2.40)**
Decision-making in monetary issues	0.813	0.818	-0.005 (-0.12)	0.026 (0.44)

Note: Figures in parentheses are t-statistics, and * and ** refer to significance levels of 10 and 5 percent respectively

Source: World Bank-AEPC survey 2009.

Independence in women's decision-making is higher in MH households than that in non-MH households (table 4.7). About 84 percent of women from MH households have a voice in their fertility issues, compared to 73 percent of the women from non-MH households. According to PSM estimates, MH connectivity improves this decision-making independence by 4.2 percentage points. MH connectivity also improves women's decision-making in issues related to their children by 2.7 percentage points, while having no impacts on monetary decision-making by women.

MH and climate change

A significant portion of CO_2 emissions can be eliminated by switching from traditional to modern fuels. For example, 90 percent of the rural population of the developing countries consumes about 730 million tons of biomass fuel every year, emitting about one billion tons of CO_2 into the atmosphere, which accounts for about 5 percent of the global CO_2 emissions (WHO 2006; Yevich and Logan 2003). In this section we take a look at the issue of CO_2 emission in the context of MH use by rural households of Nepal, that is, how much CO_2 emission the households can possibly reduce by switching to MH.[2] True, biomass consumption emits a significant quantity of CO_2, but biomass is not what households substitute when they switch to MH in rural Nepal, and so our discussion of domestic CO_2 emission considers that is due to kerosene burning only.

MH households save on GHG emissions by replacing kerosene. The difference between the amount of kerosene the MH users used to consume before adopting MH and their current kerosene consumption can be considered the reduction or savings in their kerosene consumption due to MH connectivity. That savings can be used to calculate the reduction of CO_2 emission because of MH connectivity. In reality, however, we do not have information on the past kerosene consumption of MH users. So we identify comparable groups of non-MH users and compare their kerosene consumption with that of MH users and use the difference as proxy for savings in kerosene consumption for MH users. One way to select comparable MH and non-MH households is to select them from same land group. Land is a good indicator of household wealth and it does not fluctuate in the short term—making land a suitable basis for selection of comparable households.

About 10 million kg of CO_2 is saved every year by MH households in Nepal. We assume that burning of one liter of kerosene emits 2.8 kg of CO_2. Among the non-MH households, CO_2 emission seems to go up slowly as a household's land increases, although the highest land-group households do not necessarily consume the highest quantity. Among the MH users, the pattern is roughly the same. Most importantly, the difference in emissions between MH and non-MH households is quite significant. Overall, MH households on an average emit about 3.6 kg less CO_2 than non-MH households per month. That amounts to a reduction of more than 10 million kg of CO_2 every year all for MH households in Nepal.[3] That is significant for a small country like Nepal, where only 6 percent of the rural households have MH connectivity.

Figure 4.8: CO_2 emissions by MH connectivity

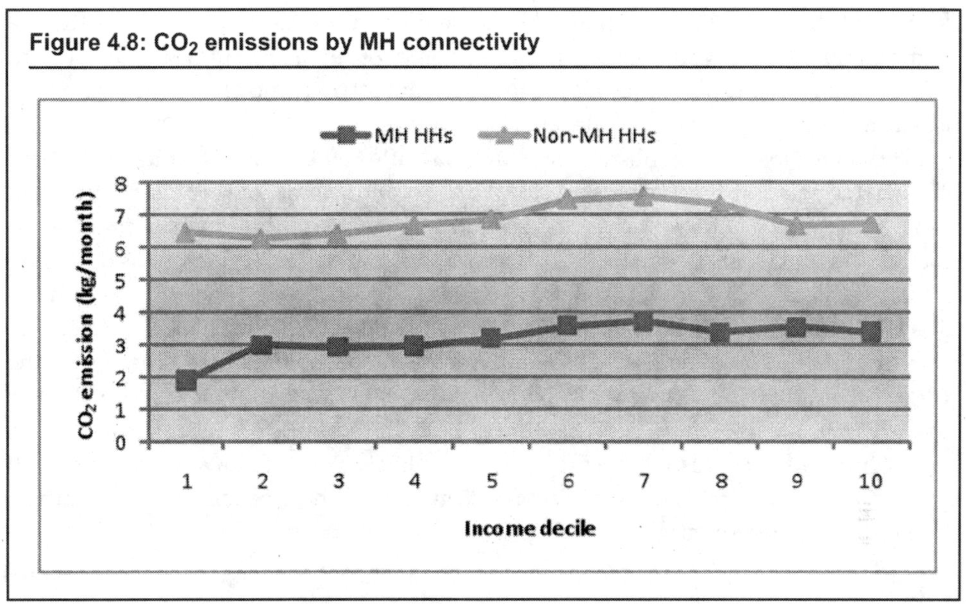

Source: World Bank-AEPC survey 2009.

Kerosene consumption of MH users can be virtually eliminated by improving the quality of MH service, which can further reduce CO_2 emissions of rural households in Nepal. MH households use kerosene as a back-up source for lighting, because of the widespread power outage and voltage fluctuations encountered by the MH users in rural Nepal. Finally, unlike other benefits that are accrued to the users of MH only (for example, income, expenditure, schooling, and health outcomes), benefits in climate change in terms of reduction in CO_2 emission are universal in their scope and duration. Many of these benefits can be achieved without making any extra investment—for example, improved and economic lighting alone justifies the switching from kerosene to MH.

Net benefits from MH electrification

The cost of delivering MH electricity to a rural household in Nepal is about Rs 6.86/kWh. This is based on a capital cost of Rs. 122,000/kW payable in 10 years with an interest rate of 12.75 percent, an operating and maintenance (O&M) cost of 5 percent of the capital cost, an average plant life of 15 years with an effective plant load factor (PLF) of 30 percent.[4] About 50 percent of this cost comes from donor and government subsidy. MH customers pay a flat tariff based on a peak power purchased, as opposed to the actual consumption in kWh per month. The tariff for MH users varies to some extent, usually from Rs 1 to 2 /w/month, and is on an average about Rs 1.5/w/month. Given the consumption pattern of the rural households in Nepal, the average effective price paid by the households is Rs 2.8/kWh which is much less than the cost incurred by the supplier

(Rs. 6.86/kWh). It clearly shows that MH intervention in rural Nepal is not cost-effective for the suppliers, even with existing subsidy. In an effort to salvage the cost, MH providers divert some of the plant power to rural micro-enterprises (mostly rice milling activities) during the day when most households do not use it.[5]

The high subsidy notwithstanding, MH may after all be a worthwhile investment when all the demand-side benefits are taken into account. First, households pay less than what they would pay for kerosene and the lighting they get is hundreds of times superior. Second, the welfare benefits (in education, income, health, and so on) as discussed in this study are also significant. Let us consider the example of consumption expenditure. The gain in household's consumption expenditure from MH use is 6.2 percent, which translates into about Rs. 420 /month for one household. On the other hand, the cost of providing MH electricity to a household is about Rs. 151/month (= Rs 6.86/kWh times the average household consumption of 22 kWh/month). That means, household's benefit from MH exceeds by its cost by about 3 times. Third, there are environmental benefits in switching from kerosene to MH. Together, all these benefits may well justify the cost of MH investments.

Notes

1. An instrument is a variable that directly influences the treatment or intervention (that is, MH connectivity) but not the outcomes – outcomes are influenced only indirectly through the intervention.
2. Although there are other GHGs (for example, methane), for simplicity, we consider here only CO_2.
3. It assumes a 6 percent MH coverage of the rural households as found from this study, a 23.5 million country population, and about 85 percent of the country population living in rural areas.
4. PwC's calculation for the World Bank.
5. That is possible because domestic use of MH electricity is mostly for lighting in the evening.

CHAPTER 5

Implementation of the Management Information System (MIS)

This chapter presents the operationalization of the M&E framework developed in Chapter 2. The framework is defined by a results chain which is translated into a set of key performance indicators to be monitored on a regular basis and a set of impact indicators that can be monitored every few years.

The true benefit of the indicators can be realized only when they are capable of being used as an organization-wide tool to monitor and track performance of its various units. Thus there is a telescopic or building-up mechanism of how information for indicators is collated and analyzed. This mechanism has been developed keeping in mind the existing management structure and M&E arrangements in AEPC.

Monitoring the performance indicators can be institutionalized in the form of a pyramid—with high-level strategic indicators focusing on outcomes and outputs at the top, facilitating decision-making process of the program managers (figure 5.1). For instance,

Figure 5.1: Build-up of information and indicators

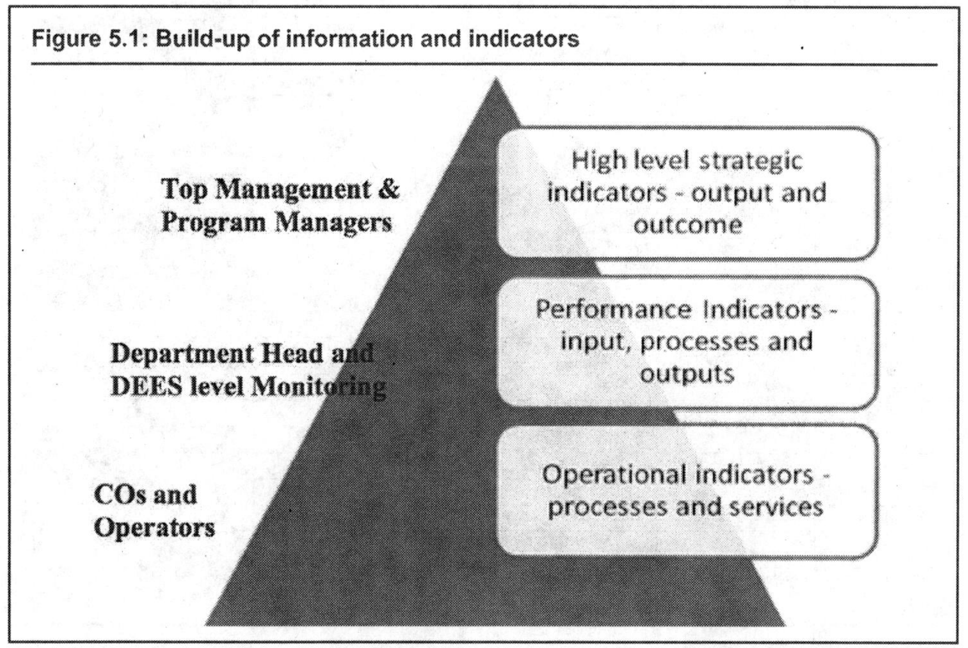

Source: Authors' elaboration.

in the case of micro-hydro, it would be appropriate to provide performance indicators, focusing on inputs, processes, and outputs for department heads and District Energy and Environment Section (DEES)-level monitoring. And at the bottom of the pyramid, i.e., the project sites, it would be appropriate to provide a larger number of operational indicators focusing on processes and services, useful for community organizations (Cos) and operators involved at the MH plant. The objective is to enable and streamline the M&E cell of REDP responsible for monitoring and evaluation of MH schemes, and to define procedures and processes to address project-related issues in an integrated manner.

The details of the indicators for each program type including the type of indicator, frequency of data collection and analysis, format reference number, and the unit responsible for collecting data is presented in Table 5.1. Reports from MIS will be generated at every level and to capture information for each KPI.

Table 5.1: Key performance indicators

Color coding used to indicate the frequency of data collection and aggregation:

Annually	
Trimesterly	
Monthly	

☑ Symbol indicates the KPIs are analyzed at the respective levels for decision making.

REDP MH Program

Sl. No.	Stage	Type of Indicator	KPIs	Unit	Monitoring Format Reference	Unit of Analysis		
						AEPC	District	CO
1	Input	Implementation Progress	No. of MH projects commissioned	No.	Format 2	☑		
2	Input	Implementation Progress	Achievement of MH installations	%	Format 2	☑		
3	Input	Implementation Progress	Progress against project's schedule	%	Format 2	☑	☑	
4	Input	Implementation Progress	Project implementation delay	Months	Format 5	☑		
5	Input	Fund Utilization	Delay in Fund Disbursement	Months	Format 2	☑		
6	Input	Fund Utilization	Utilization of Funds	%	Format 11	☑	☑	
7	Input	Fund Utilization	Variation from Budget	%	Format 11	☑	☑	
8	Input	Community Mobilization	Value of proposed equity by the community (equivalent to labor and cash amount) vis-à-vis actual contribution	%	Format 13	☑	☑	
9	Input	Community Mobilization	Percentage coverage of target population through awareness generation program	%	Format 4(a)	☑	☑	
10	Input	Capacity Building	Total no. of Training programs attended by women	%	Format 9	☑	☑	
11	Input	Capacity Building	Total no. of Training programs attended by men	%	Format 9	☑	☑	
12	Output	Technical	Installed capacity of MH	kW	Format 1	☑		
13	Output	Technical	Percentage of MH projects having skilled operators within the settlement		Format 6(a)	☑		
14	Output	Technical	No. of other RETs installed	No.	Format 20	☑		
15	Output	Technical	Achievement of other RETs installed	%	Format 20		☑	
16	Output	Technical	Design Test of MH Plant	%	Format 3		☑	
17	Output	Technical	Net generation of the Plant	%	Format 6(d)		☑	
18	Output	Technical	Distance of MH from Grid	Km	Format 1	☑		
19	Output	Technical	% of projects with Bulk metering	%	Format 4(a)	☑		
20	Output	Community Participation	Percentage of VCs became member of the MHFG	%	Format 21		☑	
21	Output	Community Participation	Percentage of households signed the agreement	%	Format 4(a)		☑	

(Table continues on next page)

Table 5.1 (continued)

22	Output	Community Participation	Number of hholds in CO/FG as % of hholds in settlement	%	Format 14		☑	
23	Output	Community Participation	Number of COs/FGs	No.	Format 21		☑	
24	Output	Community Participation	Number of FGs legalized	No.	Format 4(a)		☑	
25	Output	Community Participation	Composition of CO/FG - % Women	%	Format 21		☑	
26	Output	Community Participation	Composition of CO/FG - % VCs	%	Format 14		☑	
27	Output	Community Participation	Women in leadership positions	%	Format 4(a)		☑	
28	Output	Community Participation	VCs in leadership positions	%	Format 4(a)		☑	
29	Output	Supporting Interventions	Number of forest functional groups	No.	Format 19		☑	
30	Output	Supporting Interventions	Number of sanitation units created under the project	No.	Format 19		☑	
31	Output	Supporting Interventions	Number of nursery established	No.	Format 19		☑	
32	Output	Supporting Interventions	Number of saplings planted	No.	Format 19		☑	
33	Output	Access	Percentage of hholds with electricity access through MH	%	Format 20	☑		
34	Output	Access	No. of enterprises with access to MH	No.	Format 8	☑		
35	Output	Access	% of community facilities with access to MH	%	Format 4(a)		☑	
36	Output	Access	% of community facilities with access to other RETs	%	Format 4(a)		☑	
37	Outcome	Financial Performance	% of MHFGs with savings less than minimum saving requirement	%	Format 1		☑	
38	Outcome	Financial Performance	Percentage Investment of End User Revolving Fund	%	Format 4(b)		☑	
39	Outcome	Financial Performance	Percentage variance in cost per connection	%	Format 3		☑	
40	Outcome	Financial Performance	Percentage variance in cost per kWh of electricity generated	%	Format 3		☑	
41	Outcome	Financial Performance	Percentage O&M cost over total cost of the project	%	Format 6(d)		☑	
42	Outcome	Operational Efficiency	Percentage of Complaints resolved (consumer level) in stipulated time period	%	Format 6(b)	☑	☑	
43	Outcome	Operational Efficiency	Number of days of delay in O&M due to unavailability of spare parts at the project site	Days	Format 6(a)		☑	
44	Outcome	Operational Efficiency	Number of disconnections	No.	Format 6(c)		☑	
45	Outcome	Operational Efficiency	Number of reconnections	No.	Format 6(c)		☑	
46	Outcome	Operational Efficiency	Collection efficiency	%	Format 6(c)		☑	
47	Outcome	Operational Efficiency	Performance of Rural Energy Service Centers-Loan repaid	%			☑	
48	Outcome	Reliability	Power availability (hrs)	%	Format 6(a)		☑	
49	Outcome	Reliability	Planned supply interruptions in the month	No	Format 6(a)		☑	
50	Outcome	Reliability	Planned supply interruptions in the month	Hrs.	Format 6(a)		☑	
51	Outcome	Reliability	Unplanned supply interruptions in the month	No.	Format 6(a)		☑	
52	Outcome	Reliability	Unplanned supply interruptions in the month	Hrs.	Format 6(a)		☑	
53	Outcome	Reliability	Planned power shut downs in the month	No.	Format 6(a)		☑	
54	Outcome	Reliability	Planned power shut downs in the month	Hrs.	Format 6(a)		☑	
55	Outcome	Reliability	Unplanned power shut downs in the month	No.	Format 6(a)		☑	
56	Outcome	Reliability	Unplanned power shut downs in the month	Hrs.	Format 6(a)		☑	

Mini-Grid Support Program (MGSP)

Sl. No.	Stage	Type of Indicator	KPIs	Unit	Monitoring Format Reference	Unit of Analysis		
						AEPC	RRESC	CO
1	Input	Implementation Progress	Total no. of MH projects commissioned	No.	Format 2	☑		
2	Input	Implementation Progress	No. of MH projects commissioned at User Community Functional Group level	No.	Format 2	☑		
3	Input	Implementation Progress	No. of MH projects commissioned at Enterprises level	No.	Format 2	☑		
4	Input	Implementation Progress	Achievement of MH installations at User Community Functional Group	%	Format 2	☑		
5	Input	Implementation Progress	Achievement of MH installations at Enterprises level	%	Format 2	☑		
6	Input	Implementation Progress	Progress against project's schedule	%	Format 5	☑		
7	Input	Implementation Progress	Project implementation delay	Months	Format 5	☑		
8	Input	Fund Utilization	Delay in Fund Disbursement	Months	Format 11	☑	☑	
9	Input	Fund Utilization	Utilization of Funds	%	Format 11	☑	☑	

(Table continues on next page)

Table 5.1 (continued)

10	Input	Fund Utilization	Variation from Budget	%	Format 11	☑	☑	
11	Input	Fund Utilization	Total Expenditure of Project	Rs.	Format 11	☑		
12	Input	Community Mobilization	Value of proposed equity by the community (equivalent to labor and cash amount) vis-à-vis actual contribution	%	Format 13	☑	☑	
13	Input	Community Mobilization	Percentage coverage of target population through awareness generation program	%	Format 4(a)	☑	☑	
14	Input	Community Mobilization	Total no. of Training programs attended by women	No.	Format 9		☑	
15	Input	Community Mobilization	Total no. of Training programs attended by men	No.	Format 9		☑	
16	Output	Technical	Installed capacity of MH	kW	Format 1	☑		
17	Output	Technical	% of MH projects having skilled operators within the settlement	%	Format 6(a)		☑	
18	Output	Technical	Design Test of MH Plant	%	Format 1		☑	
19	Output	Technical	Net generation of the Plant	%	Format 6(a)		☑	
20	Output	Technical	Distance of MH from Grid	Km	Format 1	☑		
21	Output	Technical	% of projects with Bulk metering	%	Format 4(a)	☑		
22	Output	Access	Percentage of hholds with electricity access through MH	%	Format 20	☑		
23	Output	Access	No. of enterprises with access to MH	No.	Format 8	☑		
24	Output	Access	% of community facilities with access to MH	%	Format 1	☑		
25	Outcome	Financial Performance	Percentage variance in cost per connection	%	Format 3		☑	
26	Outcome	Financial Performance	Percentage variance in cost per kWh of electricity generated	%	Format 3		☑	
27	Outcome	Financial Performance	Percentage O&M cost over total cost of the project	%	Format 6(a)	☑	☑	
28	Outcome	Operational Efficiency	Percentage of Complaints resolved (consumer level) in stipulated time period	%	Format 6(b)	☑	☑	
29	Outcome	Operational Efficiency	Number of days of delay in O&M due to unavailability of spare parts at the project site	Days	Format 6(a)		☑	
30	Outcome	Operational Efficiency	Number of disconnections	No.	Format 6(c)		☑	
31	Outcome	Operational Efficiency	Number of reconnections	No.	Format 6(c)		☑	
32	Outcome	Operational Efficiency	Collection efficiency	%	Format 6(c)		☑	
33	Outcome	Operational Efficiency	Performance of Rural Energy Service Centers	%	Format 6(c)		☑	
34	Outcome	Reliability	Power availability	%	Format 6(a)		☑	
35	Outcome	Reliability	Planned supply interruptions in the month	No.	Format 6(a)		☑	
36	Outcome	Reliability	Planned supply interruptions in the month	Hrs.	Format 6(a)		☑	
37	Outcome	Reliability	Unplanned supply interruptions in the month	No.	Format 6(a)		☑	
38	Outcome	Reliability	Unplanned supply interruptions in the month	Hrs	Format 6(a)		☑	
39	Outcome	Reliability	Planned power shut downs in the month	No.	Format 6(a)		☑	
40	Outcome	Reliability	Planned power shut downs in the month	Hrs.	Format 6(a)		☑	
41	Outcome	Reliability	Unplanned power shut downs in the month	No.	Format 6(a)		☑	
42	Outcome	Reliability	Unplanned power shut downs in the month	Hrs.	Format 6(a)		☑	

Various Programs under ESAP

Biomass Program

Sl. No.	Stage	Type of Indicator	KPIs	Unit	Monitoring format reference	Unit of Analysis		
						AEPC	RRESC	LPO
1	Input	New Application	No. of Applications received & recommended for Metallic stoves	No.	Format 1	☑		
2	Input	Implementation Progress	No. of Metallic Stoves installed	No.	Format 1	☑		
3 4	Input	Implementation Progress	Achievement of Metallic Stoves installations	%	Format 1	☑		
4	Output	Capacity Building	No. of trainings conducted for Metallic Stove Manufacturers	No.	Format 1	☑		
5	Output	Operational efficiency	No. of Mud Brick Improved Cooking Stoves installed	No.	Format 2	☑		
6	Output	Operational efficiency	Achievement of Mud Brick improved Cooking Stoves installations	%	Format 2	☑		
7	Outcome	Operational efficiency	% operational of Mud Brick Improved Cooking Stoves in two years of installation	%	Format 2	☑		
8	Output	Capacity Building	No. of training programs conducted for RRESC staff- Mud Brick ICS	No.	Format 2	☑		

(Table continues on next page)

Table 5.1 (continued)

9	Output	Capacity Building	No. of training programs conducted for LPO staff -Mud Brick ICS	No.	Format 2	☑		
10	Output	Implementation Progress	No. of Institutional Improved Cooking Stoves (IICS) installed	No.	Format 3	☑		
11	Output	Implementation Progress	Achievement of Institutional Improved Cooking Stoves (IICS) installations	%	Format 3	☑		
12	Output	Capacity Building	No. of training programs conducted for Institutional Improved Cooking Stoves (IICS) promoters	No.	Format 3	☑		
13	Input	Capacity Building	No. Promotional activities conducted by RRESCs for Institutional Improved Cooking Stoves (IICS)	No.	Format 3	☑		
14	Outcome	Operational efficiency	% operational Institutional Improved Cooking Stoves (IICS) in each year of installation	%	Format 3	☑		
15	Input	Implementation Progress	No. of Applications received and recommended for Household Gasifiers	No.	Format 4	☑		
16	Output	Implementation Progress	No. of Household Gasifiers installed	No.	Format 4	☑		
17	Output	Implementation Progress	Achievement of Household Gasifiers installation	%	Format 4	☑		
18	Output	Capacity Building	No. of trainings conducted for private sector development of Household Gasifiers	No.	Format 4	☑		
19	Outcome	Operational efficiency	% Operational Household Gasifiers in a year	%	Format 4	☑		
20	Input	Implementation Progress	No. of Applications received and recommended for Institutional Gasifiers	No.	Format 4	☑		
21	Output	Implementation Progress	No. of Institutional Gasifiers installed	No.	Format 4	☑		
22	Output	Implementation Progress	Achievement of Institutional Gasifiers installations	%	Format 4	☑		
23	Output	Capacity Building	No. of trainings conducted for private sector development of Institutional Gasifiers	No.	Format 4	☑		
24	Outcome	Operational efficiency	% Operational Institutional Gasifiers in a year	%	Format 4	☑		

Solar Program

Sl. No.	Stage	Type of Indicator	KPIs	Unit	Monitoring format reference	Level of Analysis		
						AEPC	District	Company/ CO
1	Input	New Application	No. of Applications received for Institutional Solar PV System (ISPS)	No.	Format 1(b)	☑		
2	Input	New Application	No. of Application received for Solar Photovoltaic Pumping System (PVPS)	No.	Format 1(a)	☑		
3	Input	New Application	No. of Applications received for Solar Dryers	No.	Format 1(c)	☑		
4	Input	New Application	No. of Applications received for Solar Cookers	No.	Format 1(d)	☑		
5	Input	New Application	No. of Applications received for Solar Home systems (SHS)	No.	Format 1(e)	☑		
6	Input	New Application	No. of Applications approved for Small Solar Home systems (SSHS)	No.	Format 1(f)	☑		
7	Input	New Application	No. of Applications approved for ISPS installation	No.	Format 1(b)	☑		
8	Input	New Application	No. of Applications approved for Solar Dryers installation	No.	Format 1(c)	☑		
9	Input	New Application	No. of Applications approved for PVPS installation	No.	Format 1(a)	☑		
10	Input	New Application	No. of Applications approved for Solar Cookers Installation	No.	Format 1(d)	☑		
11	Output	Implementation Progress	No. of ISPS installed	No.	Format 1(b)	☑		
12	Output	Implementation Progress	No. of PVPS installed	No.	Format 1(a)	☑		
13	Output	Implementation Progress	No. of Solar Dryers installed	No.	Format 1(c)	☑		
14	Output	Implementation Progress	No. of Solar Cookers installed	No.	Format 1(d)	☑		
15	Output	Implementation Progress	No. of SHS installed	No.	Format 1(e)	☑		
16	Output	Implementation Progress	No. of SSHS installed	No.	Format 1(f)	☑		
17	Output	Implementation Progress	Achievement of ISPS installations	%	Format 1(b)	☑		
18	Output	Implementation Progress	Achievement of PVPS installations	%	Format 1(a)	☑		
19	Output	Implementation Progress	Achievement of Solar Dryers installations	%	Format 1(d)	☑		
20	Output	Implementation Progress	Achievement of Solar Cookers installations	%	Format 1(e)	☑		
21	Output	Implementation Progress	Achievement of SHS installations	%	Format 1(e)	☑		
22	Output	Implementation Progress	Achievement of SSHS installations	%	Format 1(f)	☑		
23	Output	Implementation Progress	Avg. days to install ISPS	days	Format 1(b)	☑		
24	Output	Implementation Progress	Avg. days to install PVPS	days	Format 1(a)	☑		

(Table continues on next page)

Table 5.1 (continued)

27	Output	Implementation Progress	Avg. days to install Solar Dryers	days	Format 1(c)	☑		
28	Input	Financial	Total subsidy allocation for ISPS	Rs. Lakhs	Format 6(a)	☑		
30	Input	Financial	Total subsidy allocation for PVPS	Rs Lakhs	Format 6(b)	☑		
31	Input	Financial	Total subsidy allocation for Solar Dryers	Rs. Lakhs	Format 6(c)	☑		
32	Input	Financial	Total subsidy allocation for Solar Cookers	Rs. Lakhs	Format 6(c)	☑		
33	Input	Financial	Total subsidy allocation for SHS	Rs. Lakhs	Format 6(e)	☑		
34	Input	Financial	Total Subsidy allocation for SSHS	Rs. Lakhs	Format 6(f)	☑		
35	Input	Financial	No. of PVPS whose subsidy is approved	No.	Format 6(a)	☑		
36	Input	Financial	No. of ISPS whose subsidy is approved	No.	Format 6(b)	☑		
37	Input	Financial	No. of Solar Dryers whose subsidy is approved	No.	Format 6(c)	☑		
38	Input	Financial	No. of Solar Cookers whose subsidy is approved	No.	Format 6(d)	☑		
39	Input	Financial	No. of SHS whose subsidy is approved from REF	No.	Format 6(e)	☑		
40	Input	Financial	No. of SSHS whose subsidy is approved from REF	No.	Format 6(f)	☑		
41	Output	Implementation Progress	Installed capacity of ISPS in kWp	kWp	Format 1(b)	☑		
42	Output	Implementation Progress	Installed capacity of PVPS in kWp	kWp	Format 1(a)	☑		
43	Output	Implementation Progress	Installed capacity of SHS in kWp	kWp	Format 1(e)	☑		
44	Output	Implementation Progress	Installed capacity of SSHS in kWp	kWp	Format 1(f)	☑		
45	Output	Financial	% subsidy disbursement for ISPS	%	Format 6(a)	☑		
46	Output	Financial	% subsidy disbursement for PVPS	%	Format 6(b)	☑		
47	Output	Financial	% subsidy disbursement for Solar Cooker	%	Format 6(c)	☑		
48	Output	Financial	% subsidy disbursement for Solar Dryer	%	Format 6(d)	☑		
49	Output	Financial	% subsidy disbursement for SHS	%	Format 6(e)	☑		
50	Output	Financial	% subsidy disbursement for SSHS	%	Format 6(f)	☑		
51	Output	Financial	Avg. time for subsidy disbursement for ISPS	days	Format 6(a)	☑		
52	Output	Financial	Avg. time for subsidy disbursement for PVPS	days	Format 6(b)	☑		
53	Output	Financial	Avg. time for subsidy disbursement for Solar Dryer	days	Format 6(c)	☑		
54	Output	Financial	Avg. time for subsidy disbursement for Solar Cooker	days	Format 6(d)	☑		
55	Output	Financial	Avg. time for subsidy disbursement for SSHS	days	Format 6(e)	☑		
56	Output	Financial	No. of Qualified Companies for SHS	No.	Format 6(f)	☑		
57	Output	Capacity Building	No. of Qualified Companies for ISPS/PVPS	No.	Format 3	☑		
58	Output	Capacity Building	No. of Qualified Companies for Solar Dryer & Solar Cooker	No.	Format 3	☑		
59	Output	Capacity Building	No. of Qualified Companies for SHS	No.	Format 3	☑		
60	Output	Capacity Building	No. of Qualified Companies for SSHS	No.	Format 3	☑		
61	Output	Operational Efficiency	% of ISPS installations functioning successfully at year end verification	%	Format 5	☑		
62	Output	Operational Efficiency	% of PVPS installations functioning successfully at year end verification	%	Format 5	☑		
63	Output	Operational Efficiency	% of Solar Dryer functioning successfully at year end verification	%	Format 5	☑		
64	Output	Operational Efficiency	% of Solar Cooker functioning successfully at year end verification	%	Format 4	☑		
65	Output	Operational Efficiency	% of SHS functioning successfully at year end verification	%	Format 5	☑		
66	Output	Operational Efficiency	% of SSHS functioning successfully at year end verification	%	Format 5	☑		
67	Output	Capacity Building	% Achievement of Training programs (Actual no. of staff trained vs. Target)	%	Format 2	☑		
68	Outcome	Technical	No. of complaints for ISPS	No.	Format 7(a)	☑		
69	Outcome	Technical	No. of complaints for PVPS	No.	Format 7(b)	☑		

(Table continues on next page)

Table 5.1 (continued)

70	Outcome	Technical	No. of complaints for Solar Dryer	No.	Format 7(c)	☑		
71	Outcome	Technical	No. of complaints for Solar Cooker	No.	Format 7(d)	☑		
72	Outcome	Technical	No. of complaints for SHS	No.	Format 7(e)	☑		
73	Outcome	Technical	No. of complaints for SSHS	No.	Format 7(f)	☑		
74	Outcome	Technical	% of complaints resolved in stipulated time period for ISPS	%	Format 7(a)	☑		
75	Outcome	Technical	% of complaints resolved in stipulated time period for PVPS	%	Format 7(b)	☑		
76	Outcome	Technical	% of complaints resolved in stipulated time period for Solar Dryer	%	Format 7(ac	☑		
77	Outcome	Technical	% of complaints resolved in stipulated time period for Solar Cooker	%	Format 7(d)	☑		
78	Outcome	Technical	% of complaints resolved in stipulated time period for SHS	%	Format 7(e)	☑		
79	Outcome	Technical	% of complaints resolved in stipulated time period for SSHS	%	Format 7(f)	☑		

Biogas Program

Sl. No.	Stage	Type of Indicator	KPIs	Unit	Monitoring format reference	Unit of Analysis		
						AEPC	District	Company/CO
1	Output	Implementation Progress	No. of Biogas Plants (2-6 Cub.m.) installed	No.	Format 1(a)	☑		
2	Output	Implementation Progress	Achievement of Biogas Plants (2-6 Cub.m.)	%	Format 1(a)	☑		
3	Output	Implementation Progress	No. of Biogas Plants (8-10 Cub.m.) installed	No.	Format 1(b)	☑		
4	Output	Implementation Progress	Achievement of Biogas Plants (8-10 cub.m.)	%.	Format 1(b)	☑		
5	Output	Implementation Progress	No. of Biogas Plants (15-20 cub.m.) installed	No.	Format 1(c)	☑		
6	Output	Implementation Progress	Achievement of Biogas Plants (15-20 cub.m.)	%.	Format 1(c)	☑		
7	Output	Implementation Progress	Avg. days to install Biogas Plants (2-6 cub.m.)	No.	Format 1(a)	☑		
8	Output	Implementation Progress	Avg. days to install Biogas Plants (8-10 cub.m.)	No.	Format 1(b)	☑		
9	Output	Implementation Progress	Avg. days to install Biogas Plants 15-20 cub.m.)	No.	Format 1(c)	☑		
10	Output	Capacity Building	No. of NGOs involved	No.	Format 2	☑		
11	Output	Capacity Building	No. of Companies involved	No.	Format 2	☑		
12	Output	Capacity Building	Number of accredited Promoters for Biogas Plants	No.	Format 2	☑		
13	Input	Awareness Programs	Percentage coverage of target population through awareness generation program	%	Format 2	☑		
14	Input	Financial	Total subsidy allocation for Biogas Plants (2-6 cub.m.)	Rs. Lakhs	Format 5(a)	☑		
15	Input	Financial	Total Subsidy utilization for Biogas Plants (2-6 cub.m.)	Rs. Lakhs	Format 5(a)	☑		
16	Input	Financial	Total subsidy allocation for Biogas Plants (8-10 cub.m)	Rs. Lakhs	Format 5(b)	☑		
17	Input	Financial	Total Subsidy utilization for Biogas Plants (8-10 cub.m.)	Rs. Lakhs	Format 5(b)	☑		
18	Input	Financial	Total subsidy allocation for Biogas Plants (15-20 cub.m.)	Rs. Lakhs	Format 5(c)	☑		
19	Input	Financial	Total Subsidy utilization for Biogas Plants (15-20 cub.m.)	Rs. Lakhs	Format 5(c)	☑		
20	Input	Financial	Avg. delay in subsidy disbursement	Months	Format 5(a)	☑		
21	Input	Financial	% subsidy disbursement	%	Format 5(a)	☑		
22	Output	Operational efficiency	% of successful verification of operational Biogas Plants (2-6 cub.m.) after one year of installation	%	Format 4	☑		
23	Output	Operational efficiency	% of successful verification of operational Biogas Plants (8-10 cub.m.) after one year of installation	%	Format 4	☑		
24	Output	Operational efficiency	% of successful verification of operational Biogas Plants (15-20 cub.m.) after one year of installation	%	Format 4	☑		
25	Output	Operational efficiency	% Biogas Plants not meeting test criteria	%	Format 4	☑		
26	Output	Operational efficiency	% of Complaints resolved (consumer level) in stipulated time period	%	Format 6	☑		

(Table continues on next page)

Table 5.1 (continued)

27	Output	Capacity Building	No. of staff trained of construction companies	No.	Format 2	☑		
28	Output	Capacity Building	No. of Female users trained	No.	Format 2		☑	
29	Output	Capacity Building	Training provided to NGOs	No.	Format 2		☑	
30	Output	Capacity Building	No. of Female staff employed	No.	Format 2		☑	
31	Output	Capacity Building	No. of Male staff employed	No.	Format 2		☑	
32	Output	Capacity Building	No. of companies owned by Females	No.	Format 2		☑	

Improved Water Mill Program

Sl. No.	Stage	Type of Indicator	KPIs	Unit	Monitoring format reference	Unit of Analysis		
						AEPC	District	Company/ CO
1	Input	Implementation Progress	No. of IWM (short shaft) installed	No.	Format 1(a)	☑		
2	Input	Implementation Progress	Achievement of IWM (short shaft)	%	Format 1(a)	☑		
3	Input	Implementation Progress	No. of IWM (long shaft) installed	No.	Format 1(b)	☑		
4	Input	Implementation Progress	Achievement of IWM (long shaft)	%.	Format 1(b)	☑		
5	Output	Implementation Progress	No. of IWM electrification installed	No.	Format 1(c)	☑		
6	Output	Implementation Progress	Achievement of IWM Electrification	%.	Format 1(c)	☑		
7	Output	Implementation Progress	No. of Service Centers involved	No.	Format 3	☑		
8	Output	Implementation Progress	No. of GOAs involved	%.	Format 3	☑		
9	Output	Capacity Building	No. of IWM Kit manufacturers involved	No.	Format 3	☑		
10	Output	Capacity Building	No. of IWM Owners involved	No.	Format 3	☑		
11	Output	Capacity Building	No. of IWM Program staff involved	No.	Format 3	☑		
12	Input	Financial	Total cost incurred for IWM short shaft installation	Rs. Lakhs	Format 7	☑		
13	Input	Financial	Total cost incurred for IWM long shaft installation	Rs. Lakhs	Format 7	☑		
14	Input	Financial	Total cost incurred for IWM electrification installation	Rs. Lakhs	Format 7	☑		
15	Output	Capacity Building	Total members in GOA	No.	Format 2	☑		
16	Output	Capacity Building	Total Male members (Users)	No.	Format 2	☑		
17	Output	Capacity Building	Total Female members (Users)	No.	Format 2	☑		
18	Output	Capacity Building	Number of accredited Promoters for IWM	No.	Format 3	☑		
19	Input	Financial	Total subsidy allocation for IWM	Rs. Lakhs	Format 5(a)	☑		
20	Input	Financial	Total Subsidy utilization for IWM	Rs. Lakhs	Format 5(a)	☑		
21	Input	Financial	Avg. delay in subsidy disbursement	Months	Format 5(a)	☑		
22	Input	Financial	% subsidy disbursement	%	Format 5(a)	☑		
23	Output	Operational efficiency	% of operational IWM after one year of installation	%	Format 4	☑		
24	Output	Operational efficiency	% of IWM not meeting test criteria	%	Format 4	☑		
25	Output	Operational efficiency	Percentage of Complaints resolved (consumer level) in stipulated time period	%	Format 6		☑	
26	Output	Capacity Building	No. of staff of manufacturers trained	No.	Format 6	☑		
27	Output	Capacity Building	No. of Ghatta Owners trained	No.	Format 6	☑		
28	Outcome	Capacity Building	No. of millers getting benefited	No.	Format 6	☑	☑	

Source: Authors' elaboration.

Why does AEPC need an enhanced MIS?

Designing an effective MIS is an important component of the M&E system for AEPC. A robust MIS is essential for effective implementation of the M&E framework. This would enable the middle and top management of AEPC to have timely access to reliable and accurate information for informed decision-making. It would help AEPC monitor the development and implementation of its programs against set objectives.

The existing MIS of AEPC is ineffective as it does not provide timely, accurate, or sufficient information for effective management, nor does it provide adequate information to external stakeholders. There is manual compilation and collation of MIS statements. There is the possibility of inaccuracy, as the MIS does not emanate from a computerized system. There is very little interim review and analysis of the information being received to help determine what type of corrective measures need to be taken for achieving project goals. The decentralized system planning for various programs leads to lack of integration and standardization of systems across the organization.

The development of a computerized MIS would greatly enhance AEPC's capability to extract any desired reports on project progress status, and to understand reasons for delay and for future decision-making. The system would help data to be presented in a timely and concise manner to the relevant target audience and at different levels within the implementing agency.

What are the objectives of the MIS?

AEPC would require a robust and comprehensive system to take care of existing requirements and any new requirements arising in the future. This system easily integrates with existing applications and data sources. The objective has been to leverage the AEPC's existing capability by building on the existing infrastructure and optimal utilization of the same.

The design and implementation of a comprehensive MIS is driven by the following considerations:

- An automated system is needed to enable AEPC to effectively measure, assess, and report on its performance and work toward continuous improvement.
- It provides relevant information at various decision-making levels. It caters to the information requirements of both internal and external stakeholders, e.g., community organizations, government agencies, funding agencies, etc.
- Multiple ways of viewing data and conducting analysis are supported. The user can extract reports in desired formats at any given point of time from any location. For example, any report parameter or KPI can be viewed by individual month, cumulative months, financial year, monthly trend, and month by month comparison.
- There is a single-point field-level data entry in standardized formats at the lowest level of the organization, avoiding multiplicity of data sources with minimal manual intervention. Data points are then aggregated for evaluating performance at higher levels.
- Flexibility in analysis enables comparisons between organizational units and against predefined targets and achievement figures.

What are the attributes of the MIS?

A well-designed M&E framework supported by a robust MIS called the Management Information Reporting and Monitoring System (MIRMS) is a useful tool that would enable AEPC management to track progress and demonstrate the impact of its programs. The MIS will be implemented in an integrated manner, with adequate attention to the infrastructure and resource requirements as well as the institutional strengthening required within AEPC. For implementing the MIRMS, the availability and adequacy of existing tools and software available for the programs were reviewed. Given the current information setup for AEPC, some new procedures are proposed to be put in place for gathering, collating, and analyzing the MIS information. However, these are proposed fully recognizing the existing setup of AEPC and the constraints in terms of infrastructure and resources so as to reduce any additional burden on the staff.

The approach to develop a robust MIS is discussed below.

Development of standard MIS formats

An analysis of the information needs at different levels for each program and discussions with key officials has led to the modular design of the MIS formats. The redesigned modular MIS formats have been structured to address all the programs and functional areas in AEPC. The formats for standard reports include the data to be collected, the source of data, the frequency of data collection, and the parties responsible for collecting, analyzing, reporting, and using the data. The KPI that is derived from each format is also highlighted in each of the MIS formats.

The design of the MIS and the frequency of information flow has been based on the information needs of various management levels within AEPC (figure 5.2). The focus is not only on measuring day-to-day operational performance, but also on using it as a strategic planning tool. The periodicity of review varies depending upon the level of management targeted. In the redesigned system, information is captured at the source, with an operational review at the middle management level and a strategic review at the top management level.

- Monthly monitoring of operational data is done by the field staff responsible for the respective functions.
- Trimester monitoring of performance parameters is done by the district heads or monitoring officers of the respective programs.
- Annual, trimesterly, or monthly review of the programs and strategic parameters is done by the top management for informed decision-making and reporting to the external stakeholders.

Structured and unified formats have been designed to facilitate ease in data gathering across various locations. The list of MIS formats, including the existing formats and proposed new formats, has been provided to AEPC.

Development of computerized MIRMS

A Web-based MIRMS has been developed that would serve as the primary interface for all monitoring and analysis purposes. The system will support the decision-making capability of AEPC management by generating a set of MIS reports and evaluation

Figure 5.2: Frequency of information flow

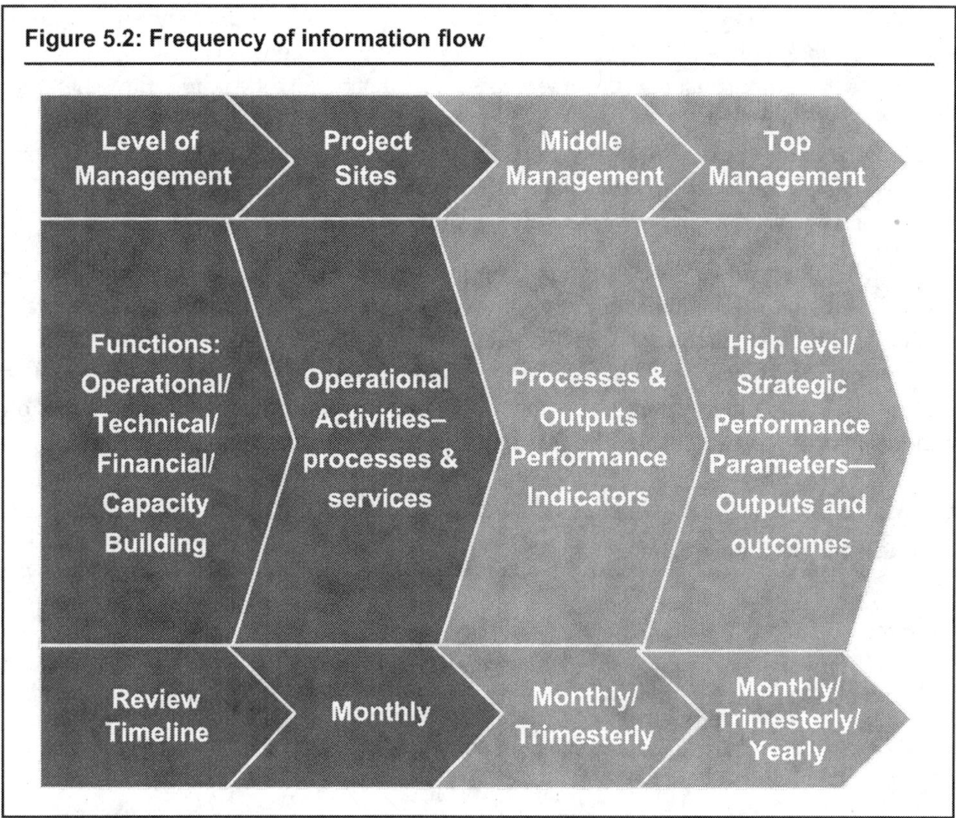

Source: Authors' elaboration.

dashboards based on monthly and quarterly program data for stakeholders. The system is designed to compile the performance parameters at the project level/VDC level and above for each of the programs.

The MIRMS will use the existing computer infrastructure of AEPC's server for hosting the system on the Internet. The system's users can access the system locally through a LAN or from any remote location through the Internet. With the implementation of the centralized AEPC server at AEPC's head office, the transaction of consolidated data from all of AEPC's programs would be populated on a periodic basis. Data would be either entered directly or extracted and uploaded in the system from different programs. The MIRMS will interface with other external systems either through an online or offline mode to extract information captured through other databases.

The MIRMS is a customized "business intelligence tool" developed on a 3-tier Web technology framework comprising of Java technology integrated with MS SQL Server 5. The MIRMS is to be implemented at AEPC server located at AEPC's office in Kathmandu. Application server Tomcat 6 would be configured in the server to host the system. A real IP would be required to host the system on the Internet.

- Database Layer—consists of database tables and stored procedures. It is compatible with both SQL Server 5 and Oracle 10g, whichever is available.
- Application/Business Layer—consists of JSP and Java Bean in which the business is embedded.
- Presentation layer—consists of JSP pages and scripting languages like JavaScript. This layer uses the client systems browser to display the front end either on a LAN, WAN, or over the Internet.

The system has been conceptualized and designed on an open architecture and extensible framework to facilitate integration with external systems and existing software available within AEPC (figure 5.3). User-friendly interfaces are being provided for mapping and uploading external data sources to MIRMS. Data migration procedures will be built in the system to interface with other databases for acquiring information either on a monthly or trimesterly basis. The frequency of data acquisition can also be configured in the system.

Figure 5.3: Data flow structure

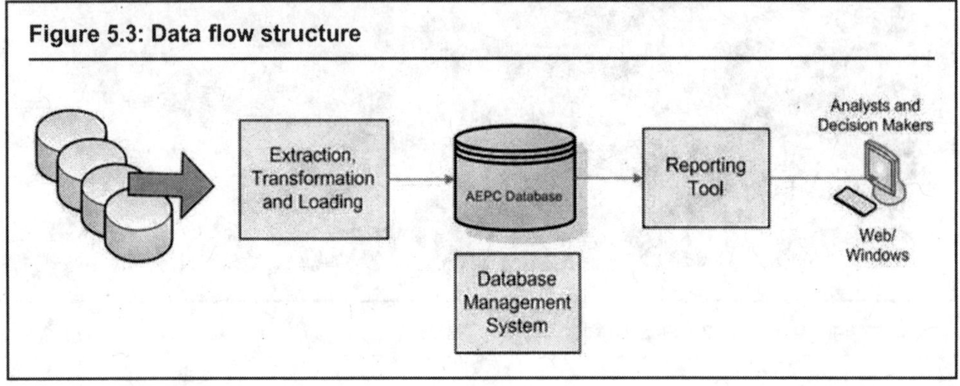

Source: Authors' elaboration.

The MIS server located within the AEPC office would be the central database repository with data integration infrastructure (figure 5.4). The infrastructure will have the capability to extract, transform and load data acquired from multiple sources including periodic progress reports, consolidated outputs from database, yearly program targets from program document, subsidy disbursement information, DPR information, and other transactional data. For REDP, the software installed at each of the district DEES offices and at the M&E cell at the head office would be used for transferring monthly information from DEES to the head office. Similarly, for ESAP, Microsoft Dynamics NAV installed at each of the district Regional Renewable Energy Service Center (RRESC) offices and at the head office would be utilized for transferring monthly transactional information from RRESC to the head office. Monthly consolidated MIS information for both Improved Water Mill and Biogas programs would be uploaded in the system. Information not captured as a part of the existing formats would be entered manually in the new prescribed formats, preferably in Excel sheets, and will be sent to the head office via the fastest and most secure mode of transmission. The new formats would then be uploaded in the AEPC server through data uploading screens. Similarly, all other information relating to different aspects of the project, such as DPR information, subsidy disbursement, program targets, etc., available either at the district or at the head office, will be inputted in prescribed MIS formats and uploaded in the AEPC server.

Figure 5.4: Institutional structure of MIS

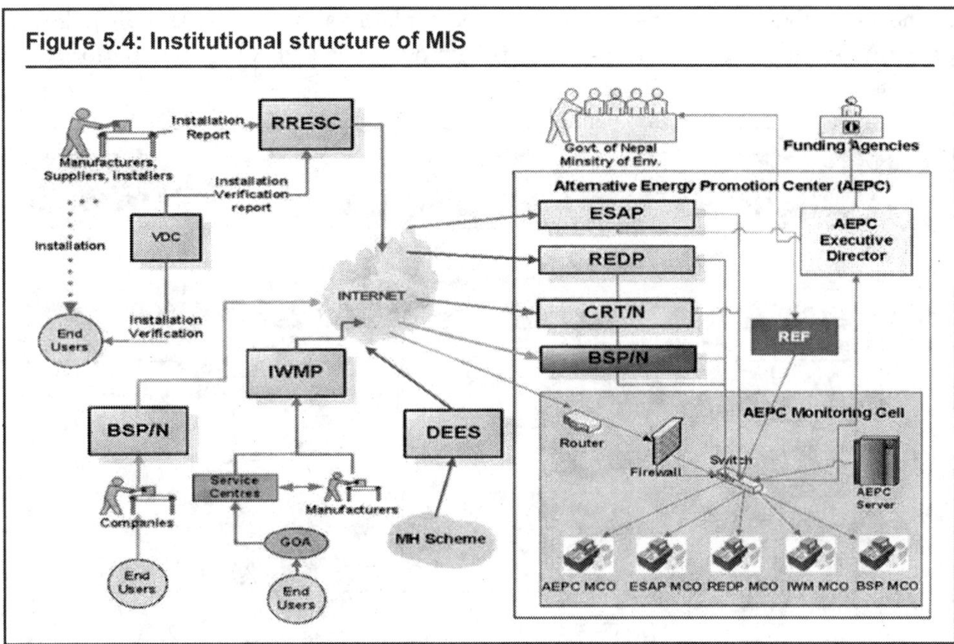

Source: Authors' elaboration.

The key features of the MIRMS are as follows:

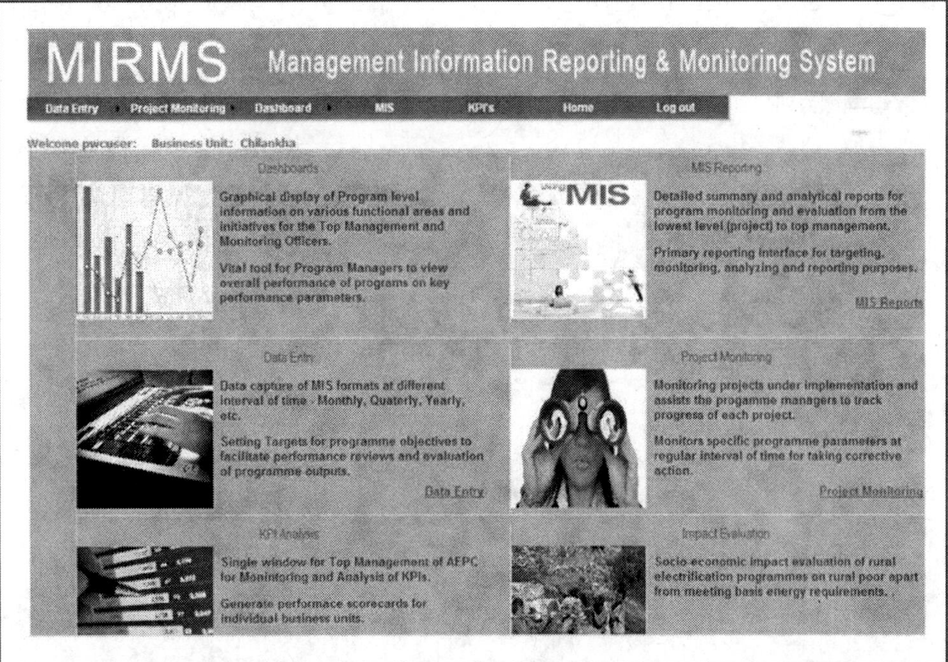

Data entry: This module provides for data entry for various programs at different stages and levels. The data entry screens have been designed keeping in mind the availability of performance data at different intervals of time. User-friendly screens allow for the mapping and uploading of an external data source to the MIRMS. Along with data entry, the MIRMS also has a provision to set targets in the system for selected MIS fields.

At the end of the desired interval, the achievement figures are compared against the target figures to arrive at the achievement of the program output, facilitating performance reviews.

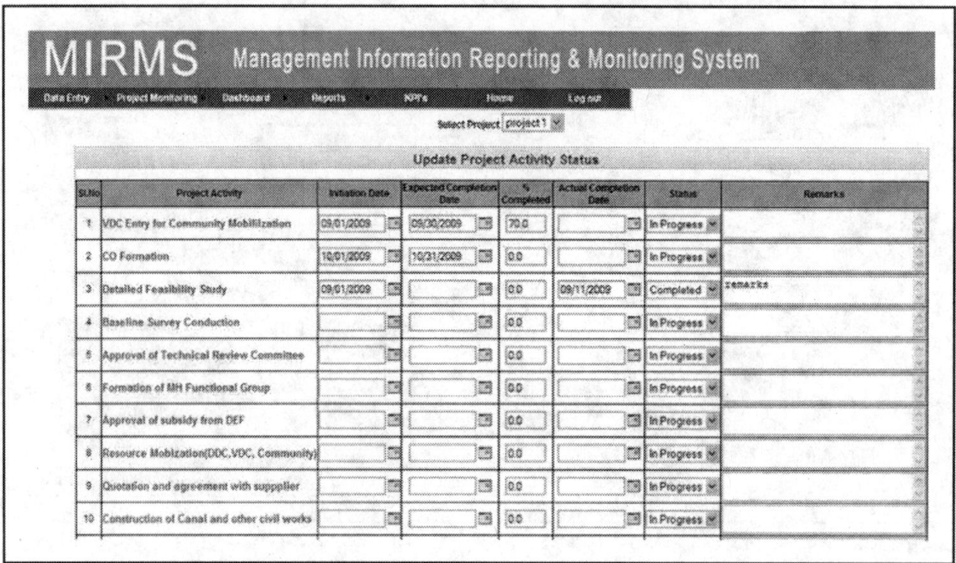

Project monitoring: To compare program/project execution performance and take corrective actions where necessary, it is necessary to monitor specific program parameters at regular intervals of time. This module tracks the project schedules for all programs under implementation and assists the program managers to track the progress of each project.

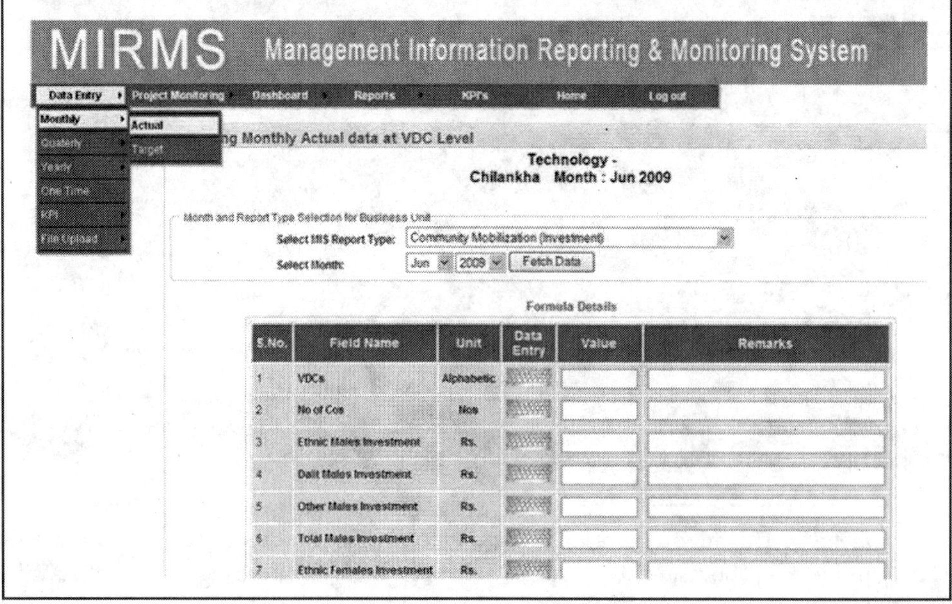

Dashboards: Features like dashboards have been defined to provide specific information about the performance of each program component and to present critical measurable parameters of each program at a glance. This gives the management and various levels in the organization a quick snapshot of the programs. The dashboards would provide a graphical display of the overall performance of the programs on key performance parameters like achievement of MH installations, community access, financial performance, etc. The evaluation mechanism would provide such dashboards on a monthly, trimesterly, and annual basis depending on the availability and frequency of data consolidation. For example, the dashboard for the project manager for REDP is presented below.

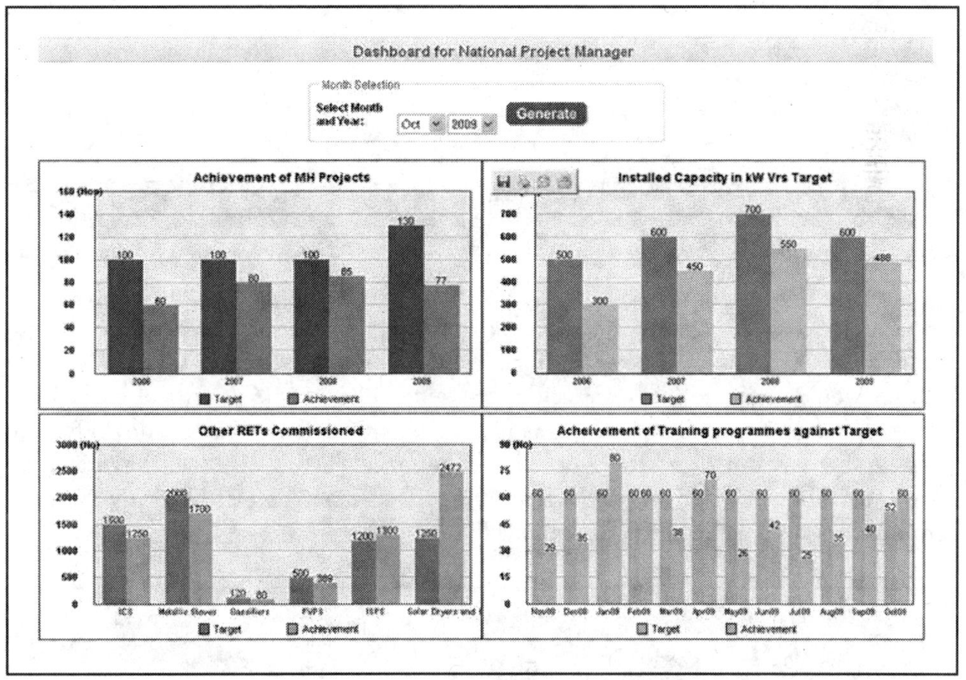

KPI analysis: The KPI analysis module provides the latest representative information to the management to support strategic activities such as goal setting, planning and forecasting, and tracking performance. It would assist the program managers in evaluating ongoing and completed projects and can be the basis for making informed decisions. The system would enable the generation of a list of high performers as well as the worst performers at different levels of the organization, and would facilitate trend analysis of the KPIs. Appropriate weightages would be given to each of the input-, output-, and outcome-level KPIs so that scorecards can be generated at all levels of the business unit.

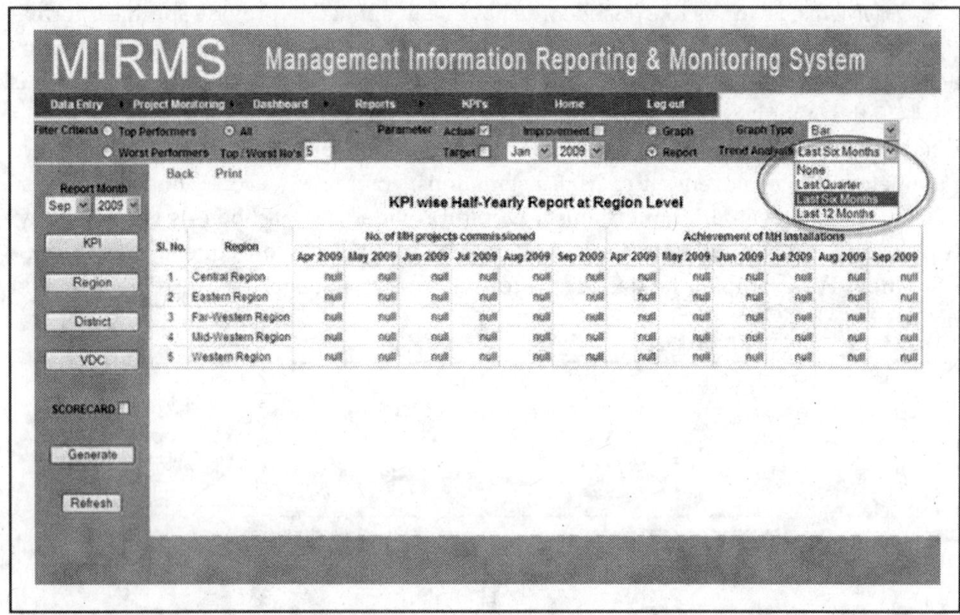

Sl. No.	Region	No. of MH projects commissioned						Achievement of MH installations					
		Apr 2009	May 2009	Jun 2009	Jul 2009	Aug 2009	Sep 2009	Apr 2009	May 2009	Jun 2009	Jul 2009	Aug 2009	Sep 2009
1	Central Region	null	null	null	null	null	null	null	null	null	null	null	null
2	Eastern Region	null	null	null	null	null	null	null	null	null	null	null	null
3	Far-Western Region	null	null	null	null	null	null	null	null	null	null	null	null
4	Mid-Western Region	null	null	null	null	null	null	null	null	null	null	null	null
5	Western Region	null	null	null	null	null	null	null	null	null	null	null	null

Detailed MIS reports: To support AEPC in its MIS reporting requirements on a regular basis, the MIRMS, through its dynamic reporting engine, provides for the generation and consolidation of MIS reports for all renewable energy technologies (RETs). The MIRMS would facilitate the M&E cell in the generation of MIS reports at every level of the organization hierarchy from time to time. The reports have a drill-down feature wherein the user can start analyzing the performance starting at the top organization level and drilling down to the last level. All the reports can be converted to any standard format and sent to the concerned locations for action.

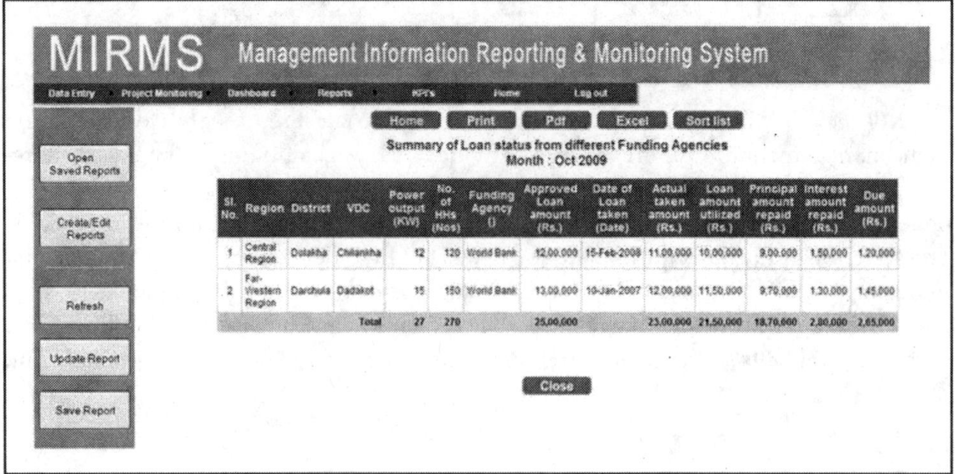

Sl. No.	Region	District	VDC	Power output (KW)	No. of HHs (Nos)	Funding Agency ()	Approved Loan amount (Rs.)	Date of Loan taken (Date)	Actual taken amount (Rs.)	Loan amount utilized (Rs.)	Principal amount repaid (Rs.)	Interest amount repaid (Rs.)	Due amount (Rs.)
1	Central Region	Dolakha	Chilankha	12	120	World Bank	12,00,000	15-Feb-2008	11,00,000	10,00,000	9,00,000	1,50,000	1,20,000
2	Far-Western Region	Darchula	Dadakot	15	150	World Bank	13,00,000	10-Jan-2007	12,00,000	11,50,000	9,70,000	1,30,000	1,45,000
			Total	27	270		25,00,000		23,00,000	21,50,000	18,70,000	2,80,000	2,65,000

How will the reporting on indicators be presented?

The following information is suggested to be reported to various internal and external stakeholders on a periodic basis to facilitate informed decision-making.

Government of Nepal

This would include reporting on the following parameters on a trimesterly or annual basis:

- Access and use of services: Percentage of households, enterprises, and community facilities with access; power availability (hrs); and quality of supply
- Financial performance: Loan Status, cost recovery, and subsidy use
- Physical outputs: Number of projects commissioned and installed capacity
- Impact of services:, Impact on income, health, education, schooling, and quality of life

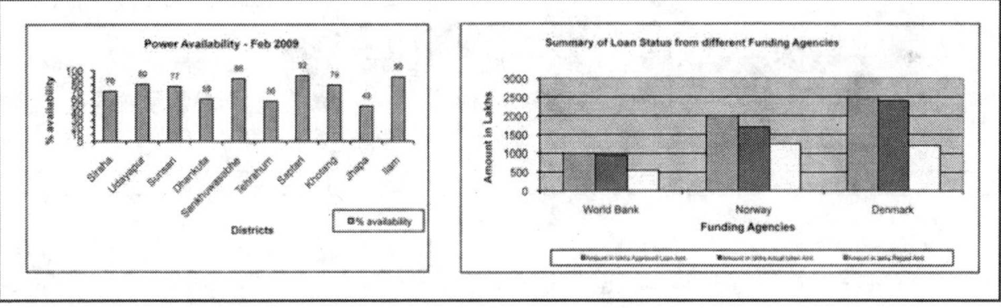

Beneficiaries

This would include reporting on the following:

- *Impact of services*: Impact on income, health, education, schooling, and quality of life—it would be useful to supplement the information on indicators with anecdotal evidence, results, and stories.
- *Financial performance*: Subsidy allocation and collection efficiency—this would be useful information to disseminate to the users of the service to encourage enhanced willingness-to-pay.

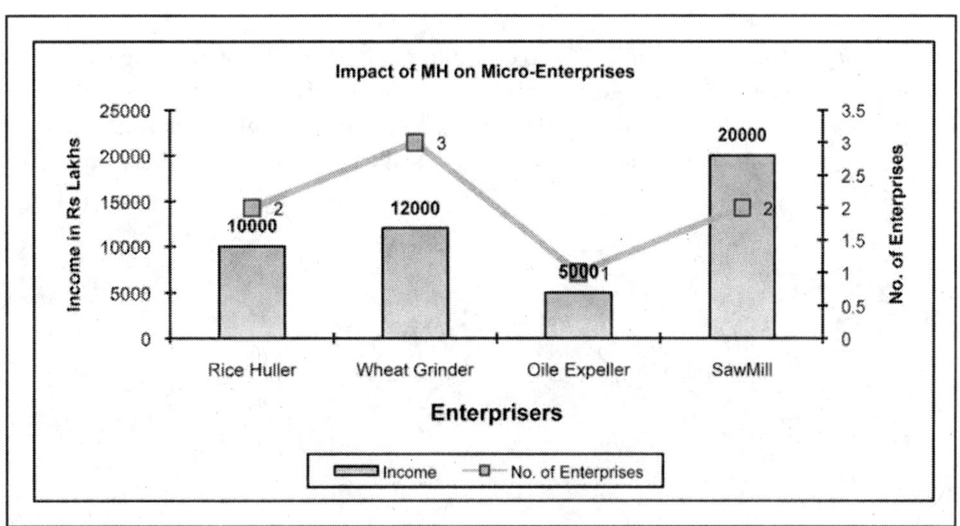

Donors/Funding Agencies

This would include reporting on the following to promote accountability by reporting on the utilization of funds, as well as to demonstrate the effectiveness by reporting on the impact of the services provided.

- *Access and use of services*: Percentage of households, enterprises, and community facilities with access; power availability (hrs); and quality of supply
- *Physical outputs*: Installed capacity and number of connections
- *Physical and financial progress*: Use of funds, percentage of subsidy disbursement, and implementation delay
- *Impact of services*: Impact on income, health, education, schooling, and quality of life

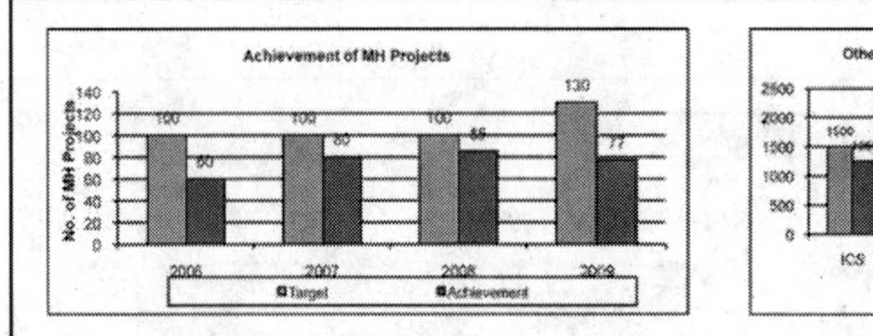

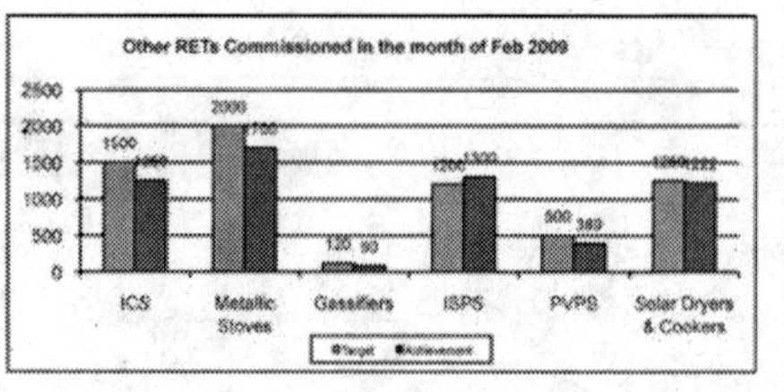

AEPC management

At the AEPC management level, the objective would be monitoring on a periodic basis to not only report to the external stakeholders but for informed decision-making and identifying any issues affecting the project's progress. This would require reporting on the following aspects.

- *Impact of services*: E.g. impact on income, health, education, schooling, quality of life
- *Access and use of services*: Percentage of households, enterprises, and community facilities with access; power availability (hrs); and quality of supply
- *Operational efficiency*: Percentage of installation meeting test criteria, collection efficiency, and complaint resolution
- *Financial performance*: Loan, cost recovery, and subsidy use
- *Physical outputs*: Number of projects commissioned and installed capacity
- *Capacity building*: Training provided and number of male and female staff
- *Community mobilization*: Population covered through awareness programs
- *Financial and physical progress*: Systems installed and implementation delay

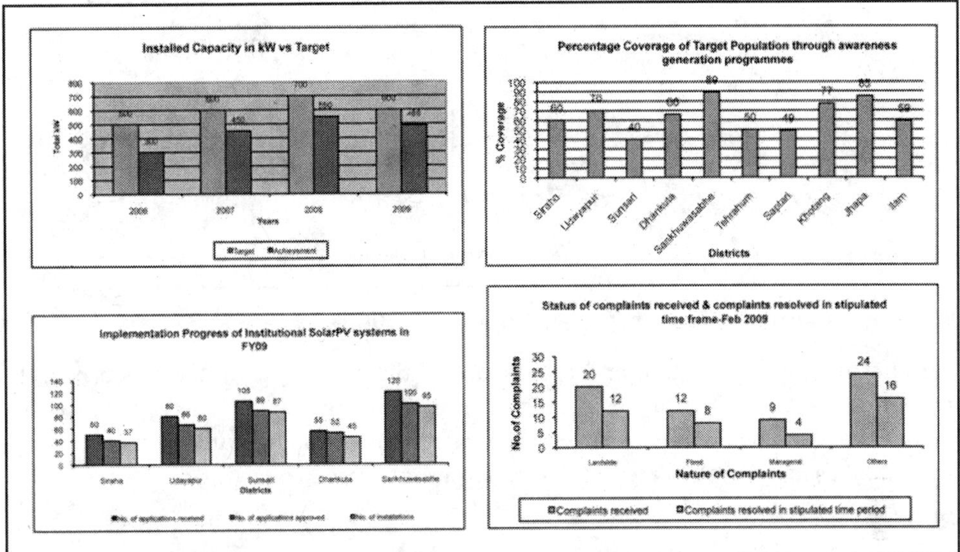

Monitoring cell, AEPC

At the AEPC Monitoring cell level, the objective would be to monitor on a periodic basis to not only report to higher level management but also to highlight any issues and take corrective action. The following aspects would be monitored on a regular basis.

- *Physical outputs*: Number of projects commissioned and installed capacity
- *Capacity building*: Training provided and number of male and female staff
- *Community mobilization*: Population covered through awareness programs and community contribution
- *Financial and physical progress*: Systems installed and implementation delay

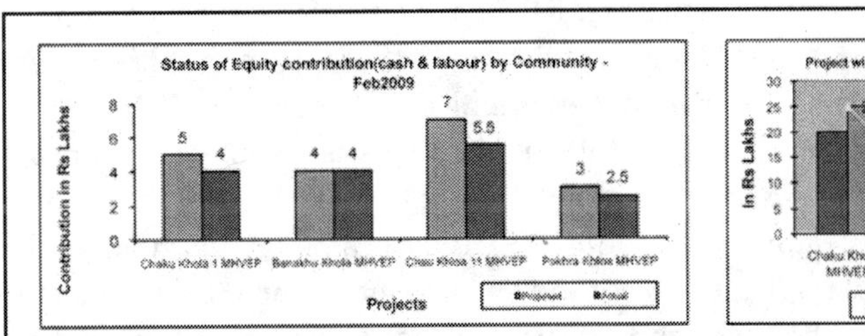

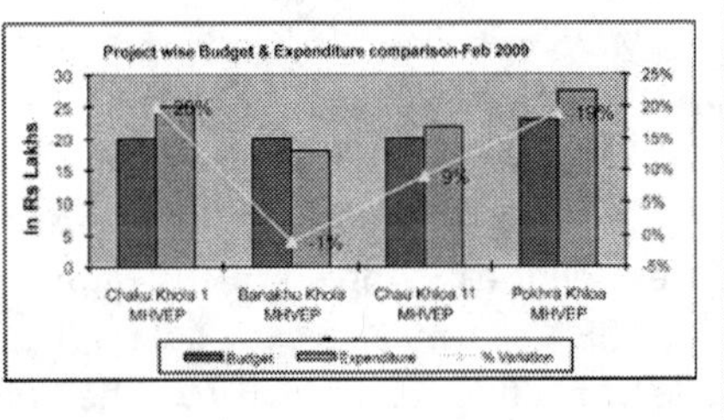

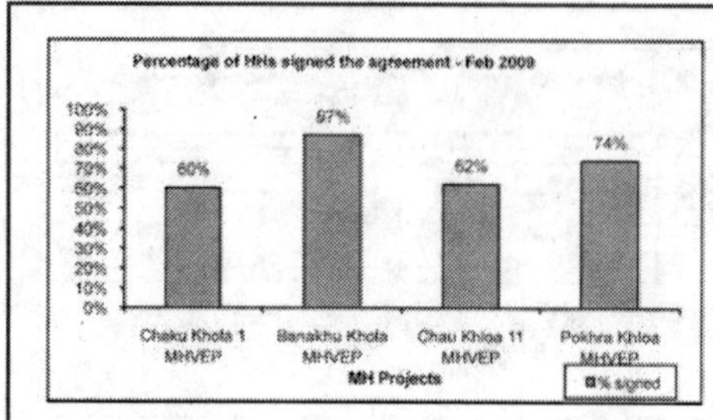

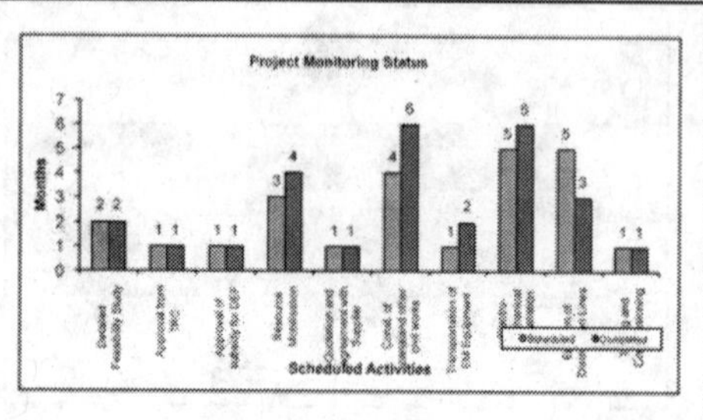

What are the institutional roles and responsibilities for the MIS?

For effective implementation and sustenance of the M&E system, there is a need to strengthen the M&E structure in AEPC in terms of the functions carried out at various levels. The M&E operational and reporting requirements along with the roles and responsibilities in the system are discussed below and presented in Annex 2.

REDP

DEES office: The EDO heads the DEES office in each district and performs the role of the program coordinator for all the MH schemes under his/her jurisdiction. At the end of the month, the community mobilizer (CM) shall record the progress for all the MH schemes located within the VDC in the prescribed formats and submit the record to the support organization (SO) located at the district headquarters. The community mobilizer coordinator (CMC), who belongs to the SO in the office of DEES, shall ensure the entering and uploading of data in the REDP software for existing formats and in prescribed Excel sheets for the new MIS formats. The EDO shall review the data before transferring it to the M&E cell of REDP. The EDO shall also be responsible for preparing the progress reports on a monthly, trimesterly, and annual basis and analyzing the data through the MIRMS to identify areas for further action. The EDO coordinates with the RRESC for collecting information on the installation of other RETs in the district.

M&E cell: The M&E cell at AEPC is staffed with two officers responsible for performing the M&E activities: (1) Monitoring & Communication Officer (MCO), and (2) Management Information Systems Associate (MISA). The MISA is responsible for coordinating and facilitating the acquisition of the monitoring data from the DEES office and other agencies involved in the implementation of REDP programs on a regular basis. The MISA also manages the data upload process and ensures that complete data is available from all program districts. The MISA should ensure the completeness of data for all the finalized MIS formats in the MIRMS system. This includes information and progress reports from the DEES office; monthly, quarterly, and annual targets set in the Annual Work Plan; and other technical, financial, and community-related specific details for each of the MH projects from the DPR once it is approved and under implementation, in addition to all other financial and operational data.

The MCO is mainly responsible for leading the M&E team to support the Program Manager and other department heads. He/she is responsible for establishing reporting needs and conducting performance analysis on a regular basis. The MIRMS would facilitate monitoring the KPIs on a regular basis to measure current performance against targets to ensure continuous improvement. A set of predefined dashboards and dynamic drill-down reports would assist the MCO in performing analysis and compiling the results into meaningful reports to be submitted to the top management for action.

RESD & RELD teams: The Regional Energy Systems Development (RESD) team includes one Senior Rural Energy System Development Advisor (SRESDA), one Rural Energy Development Advisor (REDA), and five Rural Energy Advisors, one for each development region. The team is responsible for formulating the rural energy strategy and monitoring all the MH projects from inception until commissioning. In the existing monitoring arrangement, activities scheduled under the MH projects are marked as completed on receiving the project implementation update from the EDO office, without capturing the date of completion. In the redesigned M&E framework, once the DPR is approved for implementation, for each of the activities a timeframe would be finalized and the same would be populated in the MIS server. During the execution of the project, on the completion of any activity, the date of completion would be captured and updated in the MIS server. This method of monitoring would facilitate tracking of implementation delay, if any, on a regular basis.

The Rural Energy Livelihoods Development (RELD) team includes one Livelihoods Promotion Advisor (LPA), one Gender and Social Inclusion Advisor (GSIA), and one Human Resources Development Advisor (HRDA). The team is responsible for coordinating and conducting capacity building and training programs for the DEES staff and community organizations. The MIRMS would capture the target number of training programs and other capacity building initiatives from the program document and monitor the progress of achievement. The MCO will monitor the progress of the training programs and escalate to the RELD team and top management on any slippages.

ESAP

RRESC office: The RRESC staff shall verify the data received from the LPO in given formats and enter it in the ESAP software for existing formats and in prescribed Excel sheets for the new MIS formats. The verified data is then uploaded to the AEPC server, either at the RRESC level or at the AEPC level. The staff will perform regular analysis of the progress data to ensure the implementation is on track.

M&E cell: The MCO for each of the ESAP programs—Biogas, Solar, and MGSP—will focus on performing analysis and evaluation of the data. The MIRMS would facilitate the MCO in analyzing and reviewing all programs to ensure that the programs are meeting the intended objectives. A set of predefined dashboards and dynamic drill-down reports would assist the MCO in performing analysis and compiling the results into meaningful reports to be submitted to the top management for action.

BSP

The Quality Control officer performs the quality monitoring and coordination functions for BSP. It is also recommended to appoint a dedicated MCO for the Biogas program to coordinate the MIS generation and consolidation, as well as to analyze the data on a regular basis to highlight the areas where Biogas should focus its attention for improving program effectiveness. The MIRMS would help monitor the project activities in setting up a Biogas plant, starting from a sales agreement between Biogas Companies and farmers, to the plant construction phase, to the verification process on project completion. A set of predefined dashboards and dynamic drill-down reports would assist the MCO and associates in performing analysis and recommending improvement areas in the functioning of the ongoing projects. In addition to the regular quality monitoring function, it is recommended that the Associates provide support in conducting performance analysis of ongoing projects. They should conduct field verification of all Biogas plants, and prepare the quarterly reports to submit to AEPC.

IWMP cell

Service centers: The service centers carry out social mobilization for marketing and promoting of improved water mills; study the applications of the traditional mill owners; conduct feasibility studies; coordinate with the Center for Rural Technology, Nepal (CRT/N) cell for placing orders with the manufacturers for the IWM kit and installation of the IWM; provide training and capacity building for the mill owners; and monitor system performance.

The service centers along with the coordination of the field facilitators (FFs) should perform quality control of the installations on a random sampling basis. These quality checks and subsequent monitoring visits by the CRT/N staff would ensure the quality of the IWM kits and make recommendations for enhanced service delivery. They should also conduct periodic repairs and maintenance programs for mill owners so that the IWMs run on a sustainable basis.

MCO: The MCO is responsible for coordinating the MIS generation and consolidation for the IWM program as well as for analyzing the data on a regular basis to highlight the areas where IWM should focus its attention for improving program effectiveness. For performance monitoring and measuring the achievement of program objectives, the MCO should set the targets in the system on a monthly, trimesterly, and annual basis, for all such indicators where a target is essential. The MCO should monitor the KPIs for IWM on a regular basis to measure current performance against targets to achieve improvement on continuous basis. The MCO should also establish the reporting needs of the AEPC management and coordinate with the field staff to ensure that the monthly MIS reports are consolidated and sent to the top management.

Program head

The prime responsibility of the Program Heads is to ensure that the project produces the results specified in the program document, to the required standard of quality and within the specified constraints of time and cost. The Project Head for REDP is called the National Program Manager (NPM) under the REDP, while the ESAP, at the program level, is led by the Advisor. Under the REDP, the NPM is responsible for preparing the Annual Work Plan and budget allocation based on the annual plan of District Development Committees (DDCs). The MRIMS would assist the NPM in the review of projects/ programs. The NPM is responsible for monitoring the progress of all concerned departments, reporting to him/her and other field-based institutions such as DEES and VDCs. Similarly, the ESAP's Chief Advisor is responsible for meeting reporting requirements and monitoring allied agencies under the program. BSP and IWM programmes are led by Executive Directors. The Executive Directors are responsible for meeting the reporting requirements and coordinating with AEPC for finalizing the work plan and project review. One of the prime responsibilities of the Program Heads is timely preparation and submission of various reports, both on financial and physical progress of the program, to the GoN, funding agencies, and other stakeholders of the project.

What are the risks and essential requirements?

The successful implementation of MIRMS is heavily dependent on the ready availability of relevant data from all the programs. There are certain issues that may hinder the roll out process.

- First, in the existing software currently being used by REDP and ESAP, the program level data are collected and entered at the regional and field offices. The data is then transferred to the central office either through mail or CD. The MIS formats that have been designed under the M&E framework are supposed to be mapped to the existing central database of the respective programs. The bulk of the data would be transferred to MIRMS application through data upload procedures. However, for MIS formats that are not currently being captured by the existing software, the data has to be entered directly in the MIRMS application through data entry screens. Since MIRMS is a web application, to access the application, the basic requirement is to have an internet connection. Non-existence of internet facility at few of the district offices would be a serious challenge for MIRMS roll-out in those districts. Alternative methods such as off-line data capture, emails, etc. would need to be adopted to capture data and subsequently upload at the AEPC office.
- Second, for MIS formats that have been newly introduced as part of the M&E framework, the source of data as well as data collection functionaries have to be clearly identified and necessary training should be provided to them.
- Third, it is imperative to build on the capacity of the field staff, having limited capacity pertaining to the M&E system at this stage.
- Fourth, some of the MIS formats are dependent on external agencies for data availability. The Biogas and IWM programs are run by NGOs so AEPC does not have direct access to their respective databases for monthly transfer of performance related data to MIRMS. Since, Biogas program has an Oracle database, AEPC should approach Biogas program offices for making available the required data in the prescribed MIS formats prepared as part of the new M&E framework. IWM program uses excel spreadsheet to collect and prepare their MIS and perform analysis. Necessary training should be provided to the IWM monitoring offices for using the MIRMS application to populate data and conduct periodic analysis.
- Fifth, ensuring authenticity of data that is entered by the field staff and highlighting of these data points for correction. Validation of data based on predefined logic is implemented at the data entry screens in MIRMS application, but beyond the validation limits it is difficult to prevent wrong entries or absurd values.

With the implementation of MIRMS system on the central server in AEPC, the focus of AEPC should now be directed to the following objectives to ensure long term sustainability of the system:

1. Formation of the core team comprising of representatives from each of the programs and AEPC's monitoring cell to coordinate and facilitate the information requirement of internal and external stakeholders

2. Mapping of MIRMS MIS format with existing ESAP and Biogas databases for generating the output files every month for uploading in MIRMS system
3. Roll out the system at the district and VDC level so that performance related program data is regularly uploaded in the system
4. Training of the users starting from the field level to top management on the usage of the system to carry out its M&E functions
5. Appointment of system administrator to manage the MIRMS system

AEPC needs to take immediate steps to formalize the formation of the core group members with representatives from all the programs. The core group members should receive adequate training to roll out the system for their respective programs and provide guidance and supervision to the district and field staff for effective implementation of MIRMS. AEPC has also recruited an information officer to manage the MIRMS system and to assist the ED in providing day to day information requirement extracted from the MIRMS system. The monitoring officer should coordinate with each of the program M&E staff to ensure regular flow of data to MIRMS system.

For the successful functioning of the MIRMS system and enabling timely access to information for decision making by the middle and top management of AEPC, it is essential that there is a regular inflow of program data from all the RETs to MIRMS system. The Navision Software vendor and Biogas vendor should prepare scripts in their respective software to generate the output files for uploading into MIRMS system. Procedures should be set within each programs MIS cell to regularly generate the ASCII files and upload to MIRMS system. For MIS formats that is not captured in the existing databases of ESAP and Biogas but proposed to be implemented as a part of the new M&E framework for AEPC, relevant training should be provided at the district and AEPC level for populating data in standardized format through MIRMS system.

To begin with, the MIRMS system should be rolled out in few of the selected districts for each of the programs before rolling out the system in all the districts. This initiative will enable AEPC to understand the preparedness of its district level functionaries to start using the system as well as identify any issues that may hinder the implementation of the system. The inputs and feedback obtained during the testing phase would be incorporated in the system and subsequently a full scale roll out of the system should be carried out in all the other districts.

Training and capacity building would be one of the most critical factors in ensuring sustainability of the implementation of MIRMS. Two training workshops have been conducted for the core group members comprising of the monitoring officers and program heads of each of the programs within. The training workshops were conducted to provide intensive training and build capacity within AEPC to roll out and operationalize the MIRMS application. These workshops have resulted in successful training of more than 50 officers and team members of AEPC's programs. For successful training of the district level officers, who will actually use the MIRMS application to populate data, generate reports and conduct analysis, each program should nominate officers from corporate level as well as field level for a better understanding of the system which will help ensuring sustainability of the system. The training should focus on familiarizing the users on regularly populating data for the respective MIS formats and report generation.

CHAPTER 6

Action Plan and Way Forward for AEPC

AEPC has undertaken the Strategic Organization Development Plan (SODP) initiative to redefine the mission and vision statements of the organization against its mandate through an elaborate process of stakeholders' consultation. The SODP also aims to assess AEPC's present organizational structure and propose re-engineering in the newer context. The Task Force of SODP has been active in undertaking stakeholders' consultations through workshops and meetings.

In this chapter, a diagnostic of the existing functional processes is carried out to arrive at recommendations designed to enable AEPC perform its role. These recommendations will serve as an input the AEPC's ongoing SODP. The actions relating to the resource implications for GoN and donor-funded programs of an expansion of AEPC's human and financial resources would have to be carefully considered before any of these proposals could be implemented.

What are the vision and mission of the AEPC?

AEPC was set up with the mandate to promote the use of renewable energy technologies to meet the energy needs in rural areas of Nepal. Renewable energy is considered to be the most viable and ecologically better option to the present alternatives.

The vision of AEPC is to become *"an institution recognised as a regional/international example of promoting large-scale use of renewable energy sustainably and a national focal point for resource mobilisation."* The mission of AEPC is to *"mainstream renewable energy resource through increased access, knowledge and adaptability contributing for the improved living conditions of people in Nepal."*

The 2006 Rural Energy Policy identifies the following important tasks for AEPC to promote rural energy in the country—Formulation of rural energy policy; Undertake research and studies; Formation of Rural Energy Central coordination committee under the National Planning Commission; Creation of Central Rural Energy Fund—extension of the existing REF with fund support from the GoN and other sources; Encourage formation of local groups and community based initiatives; Awareness generation and capacity building; Facilitate financial arrangement for the sector; Technical assistance to various players in the sector; Integration of RE program with other development programs; Resource assessment, especially for Solar and wind; Measures to promote manufacturing within the country; Better management and utilization of subsidy.

Analysis of the vision statement from the perspective of the tasks assigned under the Rural Energy Policy, 2006 reveals that the statement covers almost all the aspects that it has to undertake being the nodal agency for the promotion of renewable energy. Nonetheless, we understand that the statement requires revision due to the following reasons:

- The vision and mission statements were drafted way back in the year 2001. The energy context in Nepal has changed since then and there have been interventions in various forms in the field of renewable energy. Some of the technologies like MH and solar PV standalone technologies have moved up on the product maturity curve and thus these have out of the age of intensive awareness generation and promotion requiring a shift in approach and focus.
- In some of the renewable energy technologies manufacturing base needs more attention.
- Further, the statements is completely silent on energy efficiency sector, which is equally important like renewable energy as one unit of energy saved is equivalent to 1.25 unit of energy produced in Nepal.

The expectation of the industry from AEPC is also important to keep in mind while defining the vision and mission statements. During the course of this study, we interacted with private sector players and tried to understand from them the renewable energy business environment in the country and their expectation from the nodal agency AEPC to support the sector. We interacted with Solar Photovoltaics Manufacturers Association and Micro Hydro Manufacturer's Association. Their expectations from AEPC are to ensure quality of product and services; formulate policies (in the form of incentive and other support) and guidelines facilitating promotion of renewable energy; build capacity of players involved in renewable energy sector; create more installers, service providers, power house operators, managers; decrease the fund disbursement and payment processing time.

The SODP team is engaged in undertaking the task of defining the mission and vision statements through an interactive process involving a large group of stakeholders. We propose that the present business plan includes the vision and mission statements that are redefined taking into consideration the mandate of the firm, guiding acts and regulations and expectation of the stakeholders in the near future. Keeping in mind the above-mentioned mandate of AEPC and expectation of the sector from the organization, the SODP Task Force re-defined the Mission statement following a well-drawn process of stakeholders' consultation and participation. The mission statement finalized by the Task Force in consultation with stakeholders is as follows:

Mission statement of AEPC:

AEPC exists to develop, coordinate, promote, disseminate and regulate the sustainable clean energy solutions for contributing to poverty reduction in Nepal

SODP Task Force redefined the Vision statement of AEPC considering the status of the organization in a time span of next 15-20 years. The vision statement finalized by the Task Force in consultation with stakeholders is as follows:

Vision of AEPC:

AEPC is the national autonomous institution on sustainable clean/renewable energy solutions in Nepal.

Goals for AEPC were also set by the Task Force which could address the following strategic issues:

- What external changes could affect AEPC?
- What could be the influence of these changes in terms of threats or opportunities?
- What changes must AEPC make to address the threats?
- What strengths can AEPC build on to take advantage of the opportunities?
- Therefore what are the strategic issues?

Goals of AEPC:

Goal 1:
Stakeholders recognize AEPC as an effective and efficient service institution for development of RE sector.

Goal 2:
Delink from the political process (Achieve autonomous status).

Goal 3:
Enhanced staff motivation and performance; AEPC is considered as the premier institution for RE professionals in Nepal.

Goal 4:
Financial viability of investments in the sector is assured and increasing.

What are the targets facing AEPC?

AEPC has a unique role to play in the development and expansion of renewable energy technologies in rural Nepal. To promote the renewable energy sector, AEPC is driven by the goals to formulate short-, medium-, and long-term policy and plan formation; assist the GoN in formulating and implementing renewable energy policy; and ensure the integration of renewable energy in the overall national energy plan and other sectoral plans.

The National Water Plan (NWP) 2005 sets a target of 25 percent as the contribution of renewable energy of the total installed capacity in the country. Further, it is also expected that the grid interactive renewable energy plants will also contribute to the Integrated Nepal Power System (INPS) target. To achieve the target set under the NWP, AEPC will require addressing some of the key barriers such as the following:

- Undertake preparatory work to promote renewable energy, such as resource assessment, R&D support, creation of knowledge database, etc.
- Address financial barriers such as high risk perceived by financial institutions and banks for the renewable based investment along with poor financial health of NEA (which will be the purchaser of electricity in the case of grid interactive plants), which restrains the development of the sector.
- Develop a broad regulatory and policy environment. AEPC would require formulating an enabling policy targeting investment in the sector. It would also require driving for creation of a conducive regulatory environment.
- Focus on grid-based renewable energy. The focus of AEPC and the donor community has been more on Decentralized Distributed Generation and stand-alone systems in Nepal. To bring grid based RE projects into the mainstream of power generation, AEPC would require promoting this segment of renewable energy.

The development of the roadmap for AEPC is based on the targets set under the NWP 2005 that proposes household connection targets by different sources of energy, and the required total generation capacity to cater to the electricity demand for people projected until the year 2027. As targeted under the NWP, 2005, Nepal will have a total hydropower capacity of about 4,000 MW by 2027, excluding exports, and more than 75 percent of all households will be provided with INPS electricity (table 6.1). The rest of the households would be served by isolated, mini- and micro-hydropower plants and other alternative energy sources.

Table 6.1: Electrification target under the NWP 2005

Year	Total MW	HH served by INPS	HH served by Micro and Small Hydro	HH served by Alternative Energy
2007	700	35%	8%	2%
2017	2,035	50%	12%	3%
2027	4,000	75%	20%	5%

Source: NWP 2005.

Optimistic scenario—Target based on the NWP 2005

If the target suggested under the NWP, 2005 is followed, there will be total installed capacity of 1320 MW of RE in the country by the year 2027. The target highlighted in the table below is based on the assumption that RE will contribute to 3% and 4% in INPS by the year 2017 and 2027 respectively. However, seeing that thus far only about 8 MW of off-grid based MH projects have been installed in the country (projects promoted by AEPC), achievement of 1000 MW by the year 2027 appears unrealistic (table 6.2).

Table 6.2: Optimistic electrification target under the NWP 2005

	2013 (Projected)	2017 (Projected)	2027 (Projected)
Grid total capacity target (MW)	27	47	120
Off-grid total capacity target (MW)	313	417	1,000
Stand-alone systems (MW)		94	200
Total	340	558	1,320

Source: Authors' calculations based on NWP 05.

Reference case

Under this approach, the targets are disaggregated to on-grid, off-grid, and stand-alone systems, as under the optimistic scenario. For off-grid, the corresponding total installed capacity target (MW) is allocated to INPS, Micro and Small Hydro and Alternative Energy on percentile basis. After the calculation of the total installed capacity target (MW) for Micro and Small Hydro and Alternative Energy, the capacity is allocated further to different categories—Hydro capacity less than 1 MW, Solar PV, Wind, MSW and others, and Biomass. Similarly, for grid, it is assumed that the grid interactive RE plants would also add to the total generation (INPS based) of the country. Certain percentages, of the total proposed MW capacity under INPS in the NWP 2005, have been allocated to various renewable energy technologies to reach to the target for AEPC for the years 2013, 2020, and 2027. It has been assumed that the contribution of renewable energy of the total grid capacity (INPS) will be about 3 percent and 4% percent in 2020 and 2027.

The total capacity required for grid and off-grid renewable technologies is about 363 MW. On reviewing the performance of AEPC for the last 3 years in stand-alone systems, it is observed that the organization has been under-achieving the targets and there are manifold areas of improvement. As indicated in the NWP 2005, by the years 2017 and 2027, five percent of the total households is planned to be served using stand-alone systems. Against the proposed target for installed capacity in these years, the targets for stand-alone systems are 94 MW and 200 MW by the years 2017 and 2027, respectively (table 6.3).

Table 6.3: Reference electrification target[1]

	2013 (Projected)	2017 (Projected)	2027 (Projected)
Grid total capacity target (MW)	27	47	120
Off-grid total capacity target (MW)	88	131	280
Stand-alone systems (MW)		94	200
Total	115	272	600

Source: Authors' calculations based on NWP 05.

What are the actions AEPC can pursue to achieve its goals?

AEPC needs to focus on priority actions to achieve these targets. The proposed action plan is based on the diagnostic carried out as part of this study. The action plan addresses the AEPC's internal and external environment. Looking internally, the action plan reviews AEPC's existing organizational structure and proposes re-engineering. Further, the attempt has been to detail the fund requirement for the sector and the ways to address the requirement by AEPC.

Box 6.1: The core strategies of AEPC

1. Short-, medium-, and long-term policy and plan formulation:
   a. To assist HMG/N to formulate and implement a renewable energy policy.
   b. To ensure integration of renewable energy in the overall national energy plan and other sectoral plans.
2. RET development programs:
   a. To promote renewable energy technology.
   b. To help ensure supply of basic energy needs and substitute traditional energy use.
3. Standardization, quality assurance, and monitoring:
   a. To establish standards of RET and RET programs.
   b. To monitor effectiveness of RET and RET program in achieving national development goals.
4. Services and supports:
   a. To advise and assist local government and civil society in formulating and implementing renewable energy policy for enabling people to plan, implement, and manage rural energy for sustainable development through decentralised energy management.
   b. To establish a National Information Center on Renewable Energy Technology and function as a focal point of RET information network.
   c. To provide technical services for development of cost-effective and environmentally friendly energy options to private sector, NGOs, local government bodies, and civil society.
5. Subsidy and financial assistance:
   a. To institutionalize Interim Energy Fund into a full-fledged Rural Energy Fund.
   b. To provide and facilitate financial support for RET-related Research and Development.
6. Strengthening AEPC and its partners:
   a. To reorganize AEPC to reflect the service orientation of its mission statement
   b. To strengthen AEPC's/partners' organizational human resources to enable impart its responsibilities.
   c. To improve AEPC's/partners' organizational infrastructure to facilitate staff to perform to achieve its long-term vision and goals.

As is evident from AEPC's strategies, the organization's key result areas are promotion, policy setting, formulation of quality standards, support to R&D, and monitoring and evaluation in renewable energy. Implementation of renewable energy projects is not one of its core activities. AEPC would always act as a facilitator to the implementers of projects and programs in RE sector.

Source: SODP 2004.

Action Item 1: Prioritize the development of renewable technologies

Nepal is at the early stage of the renewable energy development path. The country is estimated to have robust renewable energy potential for grid and off-grid–based generation potential. However, a detailed renewable energy potential assessment is still due, and it is important that priority technologies are identified. Identification of thrust areas and setting of targets enable optimal utilization of available resources. Some of the measures for prioritization of technologies can be undertaken on the basis of the availability of renewable energy sources, adequate number of equipment suppliers for the RE technology, ease of harnessing the potential, technological advancement, and financing and institutional arrangement, etc.

Nepal can promote investment in the following areas in the near future based on a preliminary diagnostic of renewable energy:

Small/mini/micro-hydro power (grid/off-grid)

Hydro contributes about 90 percent of total energy volume in the country. The development of mega-hydropower projects over the current and future plan periods is likely to see an even greater prominence of the electricity sector within the domestic economy. This will result in improved project management skills, reduced risk, an enhanced evacuation network in remote areas, and overall development of the hydro sector.

Against this backdrop, small/mini/micro-hydro power can be a major focus area for Nepal. Investors of large hydro projects may invest in small hydro capacities near the large hydro power projects to leverage the existing infrastructure. Further, Nepal has had a history of IWM and small hydro projects that have resulted in the development of a local manufacturing base for equipment along with the skill set of local people for this technology.

Biomass and municipal solid waste grid/decentralized distributed generation (DDG)

More than 90 percent of the people in Nepal rely on firewood for cooking, and the country is primarily agrarian. Biomass and municipal solid waste (MSW) can create a win-win situation for the biomass, waste management, and renewable energy generation. Biomass-based decentralized distributed generation can be promoted especially through community-run systems.

Solar PV (including Stand-alone Systems)

The stand-alone–based Solar PV can be promoted to improve the electricity to rural areas especially to areas not connected to grid. The actual potential assessment for Solar PV in Nepal is ongoing at present. Solar PV-based stand-alone systems can be quite effective in electrification of remote areas.

Modern renewable energy services

AEPC can support renewable energy technologies such as improved furnaces and stoves, Biogas, and solar water heaters. Solar water heaters (SWH): The climate of the country is ideal for the application of solar water heaters. Electricity cost savings have been the major driver for solar water heaters in other countries. Promotion of SWH in Nepal is also possible because of high retail tariffs in the country, and its use will reduce the consumption of electricity. Solar water heaters have the potential of serving the grid-connected as well as the off-grid population. Financial incentives to users in the form of subsidy may facilitate extensive use of SWH. Appropriate incentives should be given for the promotion of small-scale enterprises for the production of solar thermal installations or their main components. Biogas: Forest biomass meets the largest share of domestic energy at about 91 percent, and hence has a significant contribution to energy security of the country. Improved furnaces and stoves: The focus on the improvement of rural stoves can lead to a reduction in the use of firewood, and also have beneficial effects on indoor air pollution and the health of the household members.

Action Item 2: Conduct preparatory work for AEPC for grid-connected renewable energy promotion

AEPC can focus for a few years on preparatory work with an aim to create an enabling environment for generation from renewable energy technologies, especially other than hydro-based projects. The initial phase will be for framing policies, and undertaking tasks such as proper resource assessment, R&D, and demonstration projects to show the viability of these projects in order to stabilize technologies. The subsequent phase will have the actual development of these technologies leading to supplying power to the grid. Some of these initiatives would require private participation.

Resource assessment

The commercial viability of various technologies needs to be established by AEPC to increase private participation in the sector, which can be established primarily through assessment of potential in the country for renewable generation. The REP emphasizes the need for resource assessment, especially for solar and wind technologies. The policy stresses development of a solar map and wind energy plan. Similarly, it is imperative that a country-level biomass assessment be undertaken in the country by AEPC that could help the developers to establish their investment plan in the biomass sector, facilitating biomass resource, location, and size of the plant.

AEPC should target to complete resource assessment in the country for various renewable energy sources as a priority. It is also recommended that AEPC, with the help of other relevant agencies, identifies sites feasible for micro- and small-hydro–based plants and ensures that techno-economic feasibility reports are prepared by the year 2011.

Development of demonstration projects

The renewable energy sector requires hand-holding in the form of demonstration projects (of high-risk/emerging technologies) highlighting the viability of grid interactive projects. Demonstration projects help the sector to contribute towards commercialization of new technologies, provide incentives for indigenization of equipment manufacturing, and provide evidence regarding future replication of new technologies and thus give confidence to the private players in investing in the sector. The target for demonstration Projects should be established in such a way that there is at least one project coming from each of the renewable technologies. AEPC can target at least 31 MW of generation projects to be completed in the next five years. The proposed mix of demonstration project is presented table 6.4:

Table 6.4: Target for demonstration project

Technology	MW
Solar	1
Wind	10
MSW and other	10
Biomass	10
Total	31

Source: Authors' elaboration.

After establishing the viability of these Demonstration Projects, the GoN should mandate one of the State Owned Banking institutions to cater to the financing needs of such projects. A policy initiative highlighting priority lending status for renewable energy projects (grid, mini-grid and off-grid) would help create a market for the renewable energy sector.

Action Item 3: Formulate an enabling policy for investment in renewable energy

There is limited private participation in the renewable energy space. The share of power supply from the community or private sector-owned systems is 17 percent. The GoN can create an enabling policy to attract private investment in the sector, including providing a clear payment security mechanism.

Renewable energy development has been an emerging agenda for countries to develop in a low-carbon growth path. They have adopted various policy, regulatory, and market interventions to create a robust market for renewable energy (table 6.5).

Table 6.5: Incentive review

	Feed-in tariff	RPS	Capital subsidies, grants, etc.	Investment or other tax credits	Sales tax, VAT, energy tax, excise tax reduction	Tradable RECs	Net-metering	Public investment or financing	Public competitive bidding
Australia		*	*			*		*	
France	*		*	*	*	*		*	*
Germany	*		*	*	*			*	
Poland		*	*		*			*	*
Portugal	*		*	*	*				
Spain	*		*	*				*	
U.K.		*	*		*	*			
U.S.	#	#	*	*	#	#	*	#	#
Brazil	*							*	*
China	*		*	*	*			*	*
India	#	#	*	*	*			*	*
Sri Lanka	*								
Uganda	*							*	

Source: Authors' elaboration.
#: Exists *: Does not exist.

A number of these initiatives are applicable to Nepal. From the perspective of investment by private parties in renewable energy and dissemination of renewable energy-based appliances and systems in rural and urban areas, AEPC should formulate the policy targeting the following areas in fiscal and financial incentives:

Fiscal incentives

Income tax exemption: In Nepal, hydropower schemes of up to 1 megawatt (MW) do not require any license for development and also do not have to pay any income tax on the revenue generated from such schemes. This government policy has greatly helped proliferate decentralized small- and micro-hydropower systems in the hilly and mountain districts of Nepal. Such exemptions should be extended to other forms of renewable energy as well.

Direct public investment: According to the NWP 2005, the government spending in the hydropower sector is about 30 percent of the financing needs. For accelerating growth in the renewable sector, public investment can increase across all renewable technologies depending on the Government's fiscal constraints.

Import duty exemption: The Government can propose import duty and sales tax exemption on renewable system equipment to be given to importers in its move to widen the usage of energy from renewable sources.

Tax credits: Tax incentives for renewable energy projects will improve the economics of renewable energy technologies, accelerate market adoption, and encourage the investment in renewable energy sector. The nature of tax credits should vary according to the stage of renewable energy development. The emphasis should be more on investment tax credit in the beginning, whilst most of the technologies are still in their nascent stage of development, except in the small- and micro-hydro sector. The Investment Tax Credit reduces income tax liability for tax-paying owners based on capital investment in renewable energy projects. Investment tax credits, earned when the capital equipment is placed into service, help offset upfront investments in renewable energy projects and will provide an economic incentive to develop and deploy renewable energy projects in the beginning. The investment tax credit should be offered to all renewable energy projects until the year 2015. It will be beneficial to continue investment tax credits beyond 2015 for more capital-intensive renewable energy technologies, such as solar photovoltaic systems and fuel cells and new renewable energy technologies. The Investment Tax credit should be followed by the Production Tax Credit that will reduce the income taxes of tax-paying owners of renewable energy projects based on the output of renewable energy facilities. The credit can be made applicable after the year 2015 for the projects that operate for more than 75 percent of the estimated plant load factor (PLF), Each kilowatt-hour (kWh) generated by an eligible facility will reduce the amount of income tax owed, which provides an economic incentive to develop and deploy technologies that harness renewable resources, such as wind, biomass, and geothermal energy.

Accelerated depreciation: AEPC and other policy formulation agencies in the country need to recognize that investors in renewable energy will be facing steep and unpredictable changes in the value of their assets—much larger than the expected physical life of their assets would imply. The introduction of AD provisions would both recognize this problem and help encourage increased investment in renewable energy sources. Provision of AD up to 80 percent of the investment will offer a robust incentive for the firms with healthier balance sheets and diversified business to invest into renewable energy in Nepal. This increases the after-tax profit earned by investors and, in turn, enhances the viability of such investments. When there is a broad goal of increasing investment in renewable energy, there is a strong case for introducing accelerated depreciation for renewable energy assets to reduce the incentive for firms to delay investing in such equipment in the hope it will be cheaper in the future. This is particularly the case when there is potential for increased sales in the short term to actually drive the cost reductions in the medium term.

Financial incentives

Capital subsidy: Capital subsidy helps reduces the up-front investment for the RE project and enhances the project viability. The policy measure will also help promote capacity addition.

Generation-based incentives (GBI): GBIs are designed to pay the customer for the environmental attributes associated with the actual production of the renewable system through the life of the asset instead of an initial, up-front incentive payment. The structure for GBI is such that for every kWh of energy produced, the developer is rewarded with an incentive. AEPC should target GBIs once some of the technologies are commercialized. AEPC should allow only GBIs for the technology that gets commercialized. It will ensure that the plants do generate power and contribute to the mainstream power generation of the country.

Other policy measures

Renewable energy obligation (REO) and renewable portfolio standard (RPS): The REO ensures that a minimum amount of renewable energy is included in the portfolio of the electricity resources serving a state. AEPC should recommend REO for entities producing/procuring conventional fuel-based power in the country such as NEA and independent power producers (IPPs) (based on conventional sources); open access and captive consumers could be assigned a target for a minimum percentage of the total procurement of power from renewable energy sources. This will ensure a long-term market for investors and will enrich the financing environment in the country, reducing the risk of RE investment.

The subsequent step to be taken by AEPC can be technology-wise procurement targets in the form of RPS. This will ensure the development and promotion of all RE technologies and minimize unhealthy competition amongst RE technology, especially for those technologies that are costlier at present but are have a promising future. For example, Solar PV and thermal-based technologies are costly in the present form compared to wind or biomass. If RPS is not assigned for Solar-based technologies, obligated entities (the entities eligible for REO or RPS) would always prefer power procurement from cheaper RE sources to meet their obligations.

Finalization of REO and RPS should be undertaken by AEPC in a manner that accounts for the total potential available in the country for all RE, the cost of procurement, and the net impact of these on consumers, as it is expected that the cost of REO/RPS will be ultimately borne by consumers.

Feed-in tariff: A feed-in tariff/preferential tariff is a mechanism to encourage the adoption of renewable energy through legislation. The regional or national electricity utilities buy renewable electricity at the rates approved by the regulator based on a cost plus reasonable rate of return. A feed-in tariff has the ability to "jump start" the market for RE technologies by providing long-term investment security and meeting the return to equity expectation of the investors. AEPC should ensure that a feed-in tariff for grid interactive RE generation is issued by a competent authority. AEPC should also suggest the methodology for tariff setting, ensuring that a cost plus approach is adopted.

Renewable energy fund: A review of international experience suggests that RE/clean energy funds have been quite effective in the development of the RE projects. The key characteristics of the RE fund can be to provide low-interest loans, fund high-risk projects, or provide mezzanine financing to new technologies or projects. Low-cost funds facilitate R&D projects and demonstration projects of high-risk/emerging RE technologies and support development of evacuation infrastructure, access roads, etc.

Single-window clearance for renewable energy projects: Scheduled execution of projects is necessary as it saves from time overrun and cost overrun. Developers face the problems of procurement of land, and getting clearances and approvals are the major barriers that generally delay the commissioning of projects. Mechanisms such as the Single Window Clearance for renewable energy projects and escorting services should be implemented in Nepal as well. Developers in Nepal would require hand-holding from institutions on clearances and other aspects, especially since many of them would be new entrants.

Action Item 4: Focus on manpower planning and organizational structure

Typically an organizational structure is a hierarchical concept of a subordination of entities that collaborate and contribute to serve the vision through the strategies employed. The organizational structure is based on the vision and mission of the organization that is the mandate under which these organizations have been formed. In this diagnostic, we have attempted to capture the mandate for the formation of AEPC and the proposed set of activities that the organization is supposed to undertake with optimal resource utilization. The organization has traditionally had a technology-based approach, which lacked proper coordination with long loops of decision-making processes (figure 6.1). To address the bottlenecks of previous technology based structure, AEPC has presently identified four core areas of activities, which are also the basis for formation of divisions—Policy, Planning and Resource Mobilization, Energy Promotion, M&E and Quality Control, and Administration.

Figure 6.1: The present organizational structure of AEPC

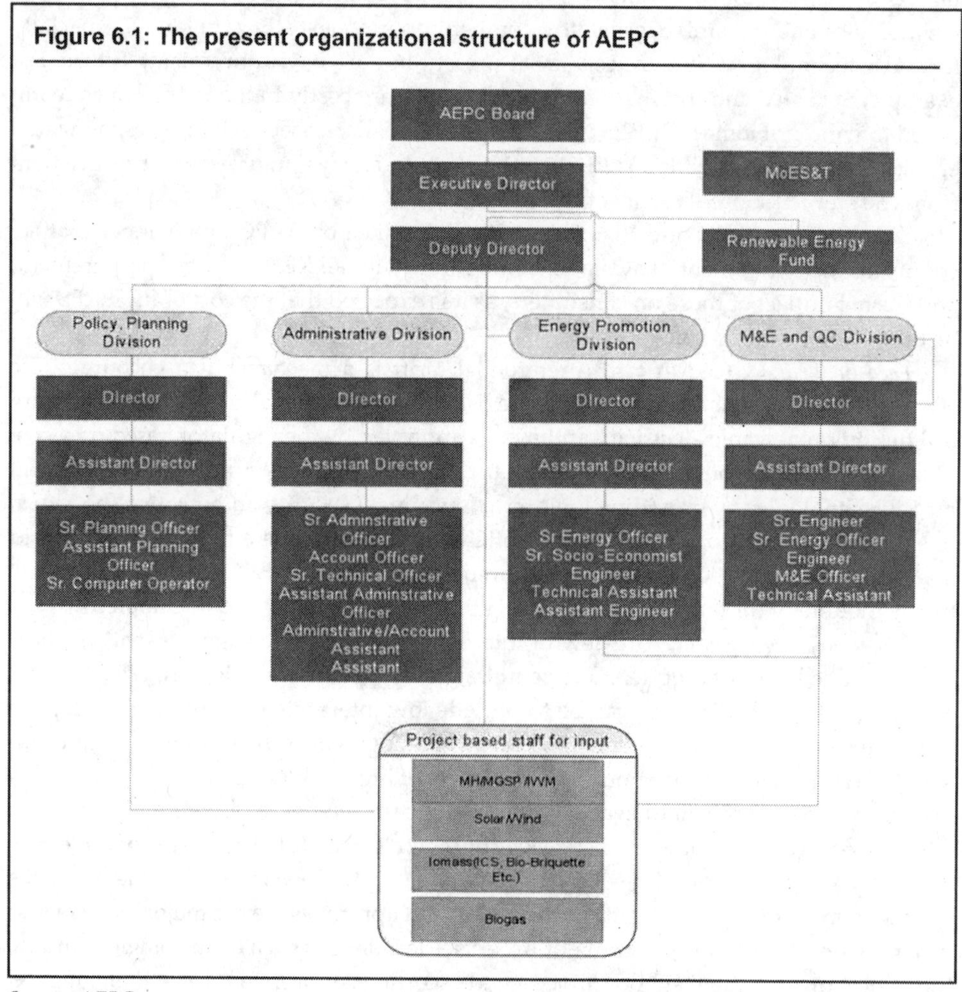

Source: AEPC.

Table 6.6: The present manpower structure of AEPC

Designation	Number	Designation	Number
Executive Director	1	Junior Officers	16
Deputy Director	1	Assistants	11
Directors	4	Drivers	3
Senior Officers	7	Peons	5
Assistant Directors	4	Sweepers	1

Source: AEPC.

The present structure has a provision of 53 staff and at present some of the positions are still vacant (table 6.6). Designations marked in yellow in the organizational structure indicate posts lying vacant at present. Against the backdrop of REP 2006 and the mandate for AEPC, strength, weakness, opportunity, threat (SWOT) analysis of the organization will provide a better analysis of the organization structure in the present context and the given mandate of the organization.

Strengths

- Recognition as a national-level apex body for RE promotion
- Rural Energy Policy in place
- Nationwide promotion of renewable energy technology
- Ensures supply of basic energy needs and is a substitute for traditional energy
- Established standards of RET and RET programs
- Advises and assists local governments on RE promotion

Weaknesses

- Potential assessment for RE sources is not yet done
- Institutionalization of Interim Energy Fund into a full-fledged Rural Energy Fund is not done so far
- Lack of direct private sector investment in grid interactive segment of RE
- R&D in selective areas only

Opportunities

- Good RE potential, especially small hydro power, to harness in the country
- Energy efficiency offers good potential
- Good IFI's fund support
- There is good rationale for the promotion of RE in the country
- Implementation of energy efficiency (EE) measures and RE applications in urban areas
- Consultancy to other firms and in other countries on RE
- Clean Development Mechanism benefits through bundling and individual projects

Threats

- Overdependence on donor support—financial stability of AEPC and the RE sector
- Promotion of RE is largely government or donor initiative
- Profit-oriented approach is lacking at present

There are areas such as R&D, policy and planning, and facilitation for private sector investment, where AEPC needs to be more active. Further, there are areas of opportunities such as energy efficiency, grid interactive RE projects, contributing to the mainstream power generation, and availability of good donor support of RE that could be leveraged by making appropriate changes in the existing organizational structure. Some of the prerequisites such as development of long-term and short-term policy on RE, potential assessment of all RE sources, and creation and institutionalization of RE Fund, are yet to be completed by AEPC. Prioritization of these measures should also be reflected in the organizational structure of AEPC.

The proposed structure is designed with an emphasis on functional areas. With a function-oriented structure, the operational processes are standardized (unlike in initial phases) and hence let the resources be used optimally, resulting in substantial cost reduction. The proposed organizational structure of AEPC consists of the following functional units: Carbon Financing, Rural Energy, Power Generation, Energy Conservation, R&D and Information Center, Policy and Planning, Finance, Administration, Central Rural Energy Fund Division and Resource Mobilization, and M&E cell.

Each division, except M&E cell and Central Rural Energy Fund, shall be headed by a Director reporting to respective Deputy Executive Directors. The Central Rural Energy Fund (CREF), which is hitherto known as Rural Energy Fund (REF), shall be under the direct supervision of the Executive Director. Directors will be assisted by Senior Engineers, Junior Engineers, and Assistants. An assessment of AEPC's manpower requirement suggests that: (i) all the RE programs would be executed under the AEPC; and (ii) the proposed organizational structure is based on the technology and programs that are implementation oriented, and thus excludes other activities of various programs such as Institutional Support to Rural Energy Sector. Based on the requirement of such programs, the organizational structure can be modified further; (iii) the present structure does not include the requirement of field-based staff, which will be based on the reach of the program and could be added as per the requirement.

Carbon financing: There is a robust potential of Clean Development Mechanism (CDM) projects for bundled stand-alone and off-grid/mini-grid–based systems. Two Biogas CDM projects have been registered by AEPC, bundling 19,396 plants. ERPA is signed between AEPC and CDCF/World Bank. We propose that a new department named Carbon Financing cater to the CDM and other similar opportunities in the country.

Rural energy division: The division will have an oversight of all the stand-alone and off-grid and mini-grid–based RE projects. The division will be headed by a Director, reporting to Deputy Executive Director. This division will also have technology-wise/program-wise Senior Engineers and a team below him consisting of Junior Engineers and Assistants, as per the requirement. The Division shall be engaged in the execution of projects, which requires presence at the ground level. We understand that most of the implementation activities will be outsourced by the Division, or services from grassroot-level organizations will be sought. However, direct staff will be required at the ground level for liaising and monitoring.

Power generation division: To meet the target set under the Water National Plan 2005, AEPC is required to take a big stride to harness the potential of grid interactive RE-based generation. To boost private sector investment in grid interactive projects, we propose a dedicated Power Generation wing to cater to the needs of the private sector, especially for grid interactive RE projects. The endeavor would require demonstration of the viability of such projects and facilitation services for private sector players for disbursement of subsidy (if applicable) or getting quick clearances. To help Policy and Planning Division undertake a country-level potential assessment for grid interactive RE generation, it shall liaison with banks and FIs to help developers for debt services. With an objective to promote grid interactive RE projects in Nepal, the Division shall work closely with the Policy and Planning Division in devising enabling policies.

Energy conservation: Energy Conservation (EC) can be one of the important measures to deal with power deficits in the country. A new division named the Energy Conservation Division is proposed within AEPC. The primary work of the Division will be to assist the Policy and Planning Division on devising Energy Conservation policies; work towards labeling programs of electric appliances; undertake energy-efficient projects; and promote energy efficiency in buildings. The Division should be entrusted to develop an Energy Efficiency road map for the country along with mapping of requirements to achieve the target. Internationally, there are international financial institutions (IFI)-supported funds available for Energy Conservation. The Division will also work closely with CREF to tap the fund. It will undertake consultancy assignments on Energy Conservation.

Research and development and information center: Special emphasis has to be given to Research & Development in RE technologies, as most of them are in the developmental stage where there is scope for improving efficiency. Adopting these new technologies requires expertise and AEPC may link up with research institutions or set them up if required. A new division needs to be set up to perform the RE technologies. The division will work closely with the Rural Energy Division on the implementation of demonstration projects. The Division is also expected to create an Information Center for Rural Energy and Energy Efficiency. The knowledge database will help investors to make informed investment decisions and will also help project executors to devise a strategy for implementation.

Figure 6.2: Proposed organizational structure of AEPC

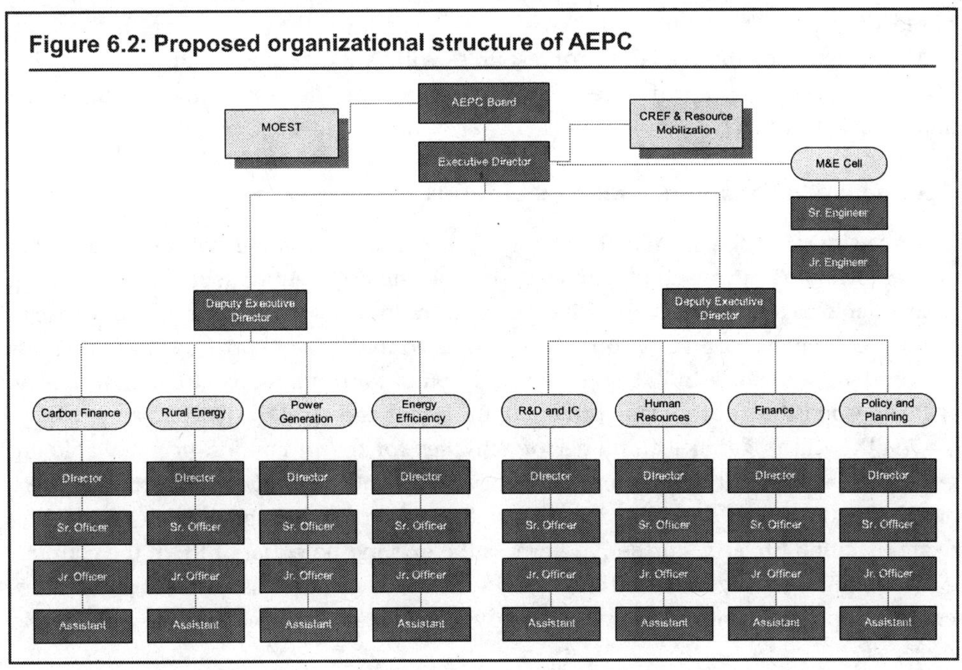

Source: Authors' elaboration.

Policy and planning division: This Division will liaise with other Divisions within AEPC and with other stakeholders to undertake the following activities in preparation for developing an overall national RE and EE plans and policies. It will work with governments on adoption of these policies and securing financial grants and coordinate with sector programs, i.e., IWM, solar, Biogas, biomass, and micro-hydro and energy conservation. Through networking with local governments, the Division would work towards the integration of national RE and EE plans into regional/local planning and proper resource mobilization. Resource assessment, both in terms of potential and commercially available renewable technologies is the prime input for the formulation of policy and subsequent RE development. This activity is not reflected under the presently identified activities by AEPC and does not find presence in the structure of the organization. The Policy and Planning Division shall take the onus of potential assessment for RE sources in the country.

Finance division: The Division will be responsible for routine finance and accounting activities, comprising receipts and payments.

Human resources division: The Division will undertake the following activities: Staff recruitment, staff training, and day-to-day administration of AEPC.

Central RE fund and resource management division: This Division is involved in assessing, raising, allocating, and managing of funds. It also needs to liaise with state agencies for subsidies and monitor their deployment by taking part in Project Appraisal.

M&E cell: Under this study, a web-based M&E framework has been developed. The M&E framework is an integrated monitoring and evaluation tool for AEPC that would help the organization to assess the performance under various programs. As the M&E is now system-based, there is no requirement for a full-fledged division, but rather a cell, reporting directly to the Executive Director. Each program or technology segment will have a Monitoring Officer, reporting to the Senior Engineer M&E Cell.

A phased approach should be adopted by AEPC to recruit manpower under proposed divisions. The number of staff shall be proportional to the shape the division takes in future, and the availability of resources with AEPC to manage the staff strength. Moreover, the appropriate number of staff under each division can further the discussed and finalized by SODP.

Action Item 5: Expand the sources of funds

The investment in the sustainable energy sector should promote not only the expansion of technology in absolute terms but also the enhancement of its impact by making it affordable to a larger section of the population. In order to add substantial capacity in different renewable energy sources, as proposed in the roadmap, huge investment is required in the sector. AEPC, being the nodal agency for promotion of RE in the country, will be responsible to attract investment in the sector and meet the targets specified.

AEPC will undertake a number of activities for the promotion of renewable energy in the country. It will require necessary funds and support for undertaking these activities. AEPC would require about Rs 1,685 million for the installation of 31 MW of Demonstration projects. Further, to harness the existing potential of RE in the country, AEPC requires to undertake various activities such as resource assessment and studies on RE, support in the form of capital subsidy to grid interactive RE projects and off-grid projects.

It is evident that AEPC's present funding sources are GoN and grant support from International Development Agencies. The GoN's contribution in FY 2007–08 was about 17 percent of AEPC's total budget, implying the dependence on International Funding agencies to route funds through programs like ESAP, REDP, BSP, and REP. It is also observed that Solar and MH programs are allocated maximum funding.

Based on the present funding allocation mechanism, we understand that for technologies that historically had a lower growth rate have been allocated lower proportions of funding (table 6.7). It is preferable that the funding allocation be done based on the future advantages like reduction in emissions, providing employment, and improving lifestyles provided by the technology.

Table 6.7: Funding source of AEPC's renewable technologies

AET	Funding by GoN (Mn. NR)	GoN's Share	Foreign Funding Sources
Solar Home Lighting System	218.78	14%	Govt of Denmark, Govt of Norway
Micro/Mini-Hydro	202.54	14%	UNDP, IDA (WB) , Govts of Denmark & Norway
Biogas	110.68	26%	Govt of Germany, Govt of Netherlands
Solar Dryer & Cooker	2.20	100%	-
Improved Water Mill	11.45	26%	Government of Netherlands
Solar Water Heating System	-	-	Private sector
Improved Cooking Stove	NA	NA	Govt of Denmark, Govt of Norway
Institutional Solar System	NA	NA	European Union

Source: AEPC.

To undertake various activities and future plans, AEPC has to arrange and generate funds on its own. There has to be some assured revenue for the organization so that it can undertake various activities for the promotion of renewable energy in Nepal.

On the basis of activities and responsibilities assigned to AEPC, the possible main sources of revenue are below.

Service fee: For assisting renewable energy-based project developers in obtaining clearances and providing administrative support for obtaining statutory clearances through clearance at various levels, the facilitation service charges at 0.5%[2] of the project cost as per the DPR that can be charged by AEPC on all commercial renewable energy projects.

Service fee for providing consultancy: Another source of income for AEPC can be through consultancy services to a varied category of clients which may include State Government, Government institutions, international and national agencies, and even private players. Other forms of income for AEPC can be through a registration fee for all the projects by private players. AEPC can assess an agency fee for managing the Green Energy Fund (GEF), which will be used in part for the promotion of RE.

Green energy fund: It is proposed that a separate GEF should be created under this policy. The different sources of generating the renewable energy fund are detailed below.

Electricity cess: An electricity cess (import or sales tax) can be one of the ways to generate the GEF. The Fund can be created by charging cess on the electricity consumed by different consumer categories in the State of Nepal. The total sales have increased, with a compound annual growth rate (CAGR) of around 6.8 percent for the period 2000 to 2009. Assuming this CAGR as the annual rate for increase of the total sales for electricity in Nepal, sales projections have been presented until 2017 (figure 6.3).

Figure 6.3: Actual and project electricity sales

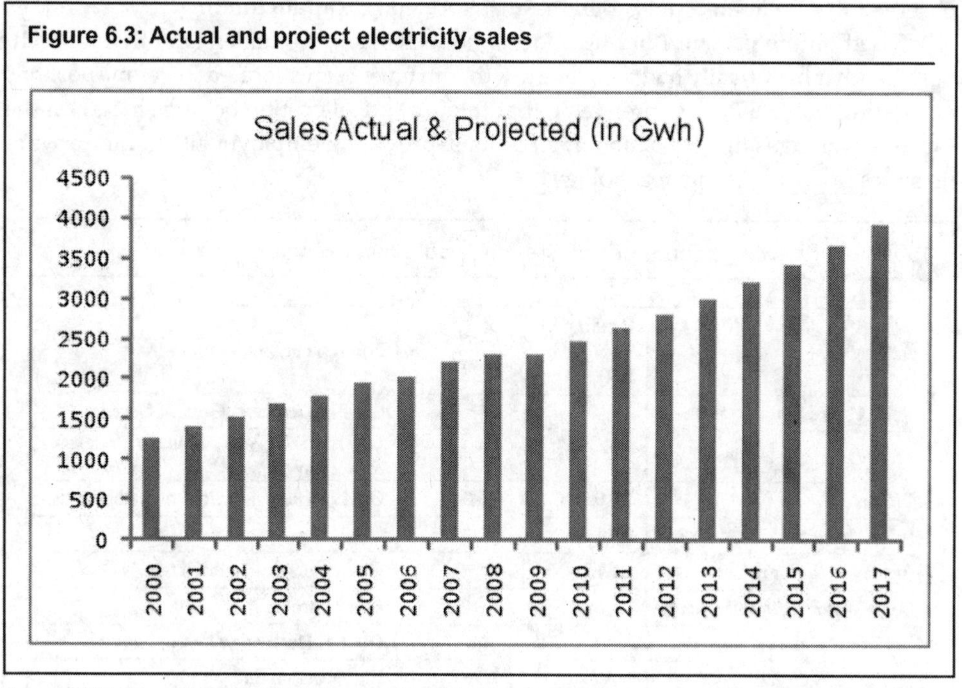

Source: NEA Annual Report (2009), Authors' calculations.

An indicative analysis to calculate the total expected collection from the cess was estimated based on the projection for the sale of electricity and assuming an electricity cess of 5/kWh on electricity sales (table 6.8).

Table 6.8: Electricity cess collection

Cess collection	2010	2011	2012	2013	2014	2015	2016	2017
Rs million	123	132	141	151	161	172	184	196
Cumulative (Rs million)								1260

Source: Authors' elaboration.

Cess on import of fossil fuel: A cess on the import of fossil fuel (petroleum products) can be an option for funding the renewable energy fund. Provision has been made under the Rural Energy Policy under the section 6.1.6 to levy tax on sales and distribution of petroleum products and utilize the fund for rural energy development.

Government support: The government can provide some type of support to GEF to promote renewable energy resources in Nepal, especially in rural areas and to promote entrepreneurship.

International agencies: Support can be sought from international agencies, international climate funds like the Strategic Climate Fund (The World Bank), the Clean Technology Fund, etc. that aim to pilot new development approaches or scale-up activities aimed at a specific climate change challenge or sectoral response, and to promote scaled-up deployment, diffusion, and transfer of clean technologies by funding low carbon programs and projects.

Clean development mechanism: Technologies that have the ability to reduce GHG emissions should be given priority, as they are eligible for CDM benefits leading to the generation of funds through carbon trading. Since these technologies are capital intensive and require huge capital, innovative mechanisms and better coordination among various parties are needed to raise funds. As in 2007, two CDM Projects of 19,396 bundled plants constructed under BSP Phase-IV have been registered with and approved by the CDM Executive Board. From these two Projects, the annual carbon revenue (net of project development and verification expenses) was Rs 42,000,000. Most of the renewable energy projects will be applicable for CDM in Nepal, as it would help reduce the generation or import of conventional fuel-based power. A certain percentage of the proceeds from CDM should be put into the GEF.

The GEF can be used for promotion of grid interactive renewable power—by means of generation based incentives, feed-in tariff, capital subsidy, etc; renewable energy for rural application and modern energy services, performance testing of projects/programs; renewable energy for urban, industrial, and commercial applications; Promotion of scientific and technological research into renewable energy; establishment of standards for the utilization of renewable energy; development of support infrastructure for renewable energy; development of demonstration renewable energy projects; conduct surveys, studies, resource assessments, and the development of relevant information systems; promotion of manufacturing of equipment for the development and utilization of renewable energy in Nepal; equity participation in renewable energy projects; energy efficiency measures in prioritized sectors; awareness generation and capacity building of energy efficiency; and promotion of programs to adopt international best practices.

From the list above, AEPC shall prioritize the activities from the list above that should be supported using the GEF. The fund should be utilized first to complete the preparatory work for the promotion of the renewable energy and energy efficiency sectors such as resource assessment, awareness generation, R&D, support for the promotion of the local manufacturing base, implementation support for energy efficiency measures in prioritized sectors.

Action Item 6: Have AEPC act as a concessional financier

The domestic financial market for renewable energy development is still nascent in Nepal. Renewable energy-based projects face financial barriers because of the high risk perceived with such projects. As a result, project developers, who are not able to obtain balance sheet-based funding, either do not get a loan or pay a higher rate of interest on debt. This has been seen as one of the factors hindering entrepreneurship in this sector.

AEPC can act as a facilitator for renewable energy projects by undertaking the role of concessional financer. Through this, AEPC could facilitate investment in renewable energy in Nepal and help project developers overcome financial barriers in raising debt. Achieving a target of around 170 MW of capacity addition by 2017 through different renewable energy technologies will require substantial debt availability of about NR 13,300 million for such projects (table 6.9).

Table 6.9: Total debt requirement

Technology	Capital cost (Rs crore/MW)	Total capacity target - grid & off-grid (MW) by 2017	Total debt required (@ 70% of total investment) (NR million)
Hydro capacity less than 1 MW	11.76	48.8	4021
Solar PV	28.56	14.1	2817
Wind	8.652	57.9	3508
MSW	10.08	17.2	1215
Biomass	7.56	32.9	1740
Total		170.9	13,300

Source: Authors' elaboration.

A substantial amount of debt is required for promoting of renewable energy projects. At times, renewable energy-based projects face financial barriers because of the high risk perceived with such projects. As a result, project developers, who are not able to obtain balance sheet-based funding, either do not get a loan or pay a higher rate of interest on debt. This has been seen as one of the factors hindering entrepreneurship in this sector. AEPC can provide loans at concessional rates to worthy project developers. This can be done by offering an interest subsidy of 2 percent to 3 percent, which AEPC would offer to the banks to lower the interest rates for developers.

Table 6.10 shows an indicative analysis for the amount of support required if an interest subsidy of 2 percent is provided for different renewable energy projects.

Table 6.10: Support required for concessional lending

Technology	Total capacity target - grid & off-grid by 2017 (MW)	Per MW interest charges for project		Per MW support required	Total support req. to meet 2017 targets
		Case 1: Base lending rate @ 12.5%	Case 2: Concessional lending rate @ 10.5%	Case 1 - Case 2 (NR million)	NR million
Hydro capacity less than 1 MW	48.8	51.45	43.22	8.23	402
Solar PV	14.1	124.95	104.96	19.99	282
Wind	57.9	37.85	31.80	6.06	351
MSW	17.2	44.10	37.04	7.06	121
Biomass	32.9	33.08	27.78	5.29	174
Total	170.9	291.43	244.80		1,330

Source: Authors' elaboration.

An amount of around NR 1,330 million is required as support to provide an interest subsidy of 2 percent and achieve a target of 170.9 MW by the year 2017. It clearly shows that if necessary support is provided to AEPC, it can play an active role to promote entrepreneurship in the green energy sector by assisting developers in getting loans at concessional rates. Such a kind of structure has the potential of providing assistance to a large number of projects. This will not only increase the investment in the sector but will also help in creating more employment.

The proposed targets will have to be revised frequently based on the following parameters. As grid-connected renewable energy is still in a nascent stage, regular monitoring of the following issues has to be done to revise the previous targets set.

- *Interim resource plan*: The assessment of potential is a continuous process, so the targets have to be revised according to the changes in achievable potential.
- *Extent of private participation*: Private participation is critical to achieve the targets, so the causal factors for changes in private participation should be studied and the targets have to be revised accordingly.
- *Advancements in technology*: Due to the developmental stage involved in various technologies, there could be significant breakthrough in technological innovations. This can help in reducing the cost of generation, which should be promoted further by increasing the targets for such technology.
- *Success of demonstration projects*: The feedback from the demonstration projects can help in revising the targets to a large extent.

Notes

1. The total investment required by 2013, 2017, and 2027 (at current cost levels) in different RE technologies for Nepal is calculated by assuming an additional cost of 5% on the capital cost of RE technologies in India. The capital cost (Rs. crore/MW) for India is assumed as Hydro capacity less than 1 MW = 7; Solar PV = 17; Wind = 5.15; MSW = 6; Biomass = 4.5.
2. The figure is indicative.

References

Barnes, Douglas F., Henry Peskin and Kevin Fitzgerld. 2003. *The Benefits of Rural Electrification in India: Implications for Education, Household Lighting, and Irrigation.* Draft paper prepared for South Asia Energy and Infrastructure, the World Bank, Washington, DC.

Barnes, Douglas F. 2008. Monitoring and evaluating impacts of rural electrification: Past experience and way forward. Energy Week Presentation, The World Bank.

Center for Rural Technology, Nepal (CRT/N). 2005. *Nepal - Focus on Renewable Energy & Poverty Reduction.* Nepal Energy Situation Survey Report, submitted to International Network for Sustainable Energy.

Independent Evaluation Group, 2008. *The welfare impact of rural electrification: A reassessment of the costs and benefits.* World Bank, Washington DC

Khandker, Shahidur, Gayatri Koolwal, and Hussain Samad. 2009. *Handbook on Impact Evaluation*, the World Bank, Washington, DC, pp:53-69.

Kulkarni, Veena and Douglas F. Barnes. 2004. *The Impact of Electricity School Participation in Rural Nicaragua*, Working Paper, University of Maryland, College Park, MD.

Ravallion, Martin. 2008. "Evaluating Anti-poverty Programs." In *Handbook of Development Economics*, vol. 4, ed. T. Paul Schultz and John Strauss, 3787–846. Amsterdam: North-Holland.

Rosenbaum, Paul R., and Donald B. Rubin. 1983. "The Central Role of the Propensity Score in Observational Studies for Causal Effects." *Biometrika* 70 (1): 41–55.

United Nations. 2005. *Energy Services for the Millennium Development Goals.* Joint publication of UN Millennium Project, United Nations Development Program (UNDP).

Winrock, 2006. Report on assessment of Rural Energy Development Program: Impacts and its contribution in achieving MDGs.

WHO (World Health Organization), 2006. *Fuel for Life: Household Energy and Health.* World Health Organization, Geneva.

World Bank, and Energy Sector Management Assistance Program (ESMAP). Washington, DC and New York: World Bank/ESMAP and UNDP.

World Bank. 2008. *The Welfare Impact of Rural Electrification: A Reassessment of the Costs and Benefits*, IEG Impact Evaluation Report, Washington, DC.

Yevich, R. and J. A. Logan, 2003. *An assesment of biofuel use and burning of agricultural waste in the developing world*, Global Biogeochem. Cycles, 17(4), 1095

Various data sources

Central Bureau of Statistics
Population of Nepal, 2002 (different volumes)
Population Monograph of Nepal (volumes 1 & 2), 2003
Population Census results in Gender Perspective (volumes 1, 2 and 3), 2002
Women in Nepal—Some Statistical Facts, 2003
National Sample Census of Agriculture, 2001/02
National Water Plan, 2005
NEA Annual Report, 2009
Statistical Year Book of Nepal, 2003
A Handbook of Environment Statistics
District level indicator for Nepal for Monitoring Overall Development, 2003
CADEC: Renewable Energy data of Nepal, 2003
Department of Health Services: Annual Report 2001/ 02
Department of Education: School level Educational Statistics of Nepal
Department of Forestry: Community Forest Bulletin no. 11
Guidelines for detailed feasibility studies of Micro-hydro projects- ESAP
REDP project documents for phase I, II and III
Guidelines for preliminary feasibility studies of Micro-hydro projects- ESAP
Summary of District Energy Study Reports—Rural Energy Development Program August, 2007
Information on Daunne Khola Micro Hydro Demonstration Scheme, September 2003

ANNEX 1:

Overview of existing M&E programs in AEPC

An assessment was done of the existing data and processes, the adequacy of existing data sources, and the roles and responsibilities of AEPC and its counterparts at the district, village, and community levels. Regular interactions were held with officials from the GoN, AEPC, donor agencies, solar and MH manufacturers associations, local beneficiaries, and other key stakeholders to have a better understanding of various expectations, the barriers to M&E, and opportunities to strengthen it.

A review of the existing M&E functions of REDP, ESAP, Biogas, and Improved Water Mill programs in AEPC was undertaken. This included a review of the existing reporting requirements, data collected to analyze performance, roles and responsibilities of key staff involved, and the supporting tools and systems. Each programmer has his/her own M&E operational and reporting requirements. A brief overview is presented in the rest of this annex.

In recent years, AEPC management has strengthened the framework for monitoring renewable energy programs. There has been an enhanced focus on monitoring various activities based on output indicators and anecdotal evidence. However, in the present scenario, there are various substantive aspects in need of improvement. These include the following:

The existing monitoring system reports on the implementation of activities but does not sufficiently address questions of access, use and quality of services, affordability, and sustainability. The current M&E system is designed to provide information on administrative, implementation, and management issues as opposed to broader development effectiveness issues and for evaluating the contribution of the programs to longer-term objectives. In the case of measuring development impact, at times it fails to incorporate the multiple stakeholders' perspectives, especially the perspectives of the beneficiaries. There is a need for systematic reporting on outputs, outcomes, and progress towards achievement of long-term objectives.

There is hardly any systematic quantification of welfare benefits. Anecdotal evidence reveals that the program has had an extremely positive and significant impact on the lives of the poor, particularly on women—reflected in health and education benefits, and income-generation opportunities. However, it is unable to measure the specific impacts on the beneficiaries in terms of improvement in health, education, income, gender empowerment, etc. Much of information is also reported through anecdotal evidence or through qualitative indicators. There is a need to capture both qualitative and quantitative information with a focus on perceptions of change among stakeholders.

There is no consolidation of information at AEPC's management level to better understand the trends and performance to support decision-making. There is no system to track information at the AEPC level by aggregating the contribution from REDP, ESAP, IWM, and Biogas programs. Each program has its unique M&E reporting requirements and systems in place. Even within the REDP program, certain indicators are manually calculated and reported at the Energy Development Officer (EDO) level, making it hard to aggregate at the central headquarters level due to absence of underlying data. There is limited capacity to perform periodic analysis and monitoring for better understanding and informed decision-making. The information collected by AEPC is not being analyzed within the M&E cell at AEPC on a systematic basis. Even though external consultants have been hired to analyze the data and generate reports based on demand from time to time, it is not done on a regular basis. Within REDP, there is only one person to manage M&E activities that are limited to data collection and record keeping. There is a need for a centralized M&E system within AEPC to facilitate a common purview of these unique programs and greater clarity in roles and responsibilities for M&E. Moreover, implementation of the computerized MIS would greatly enhance the capability of the organization for extracting any desired reports on project progress status, understanding reasons for delay and for future decision-making.

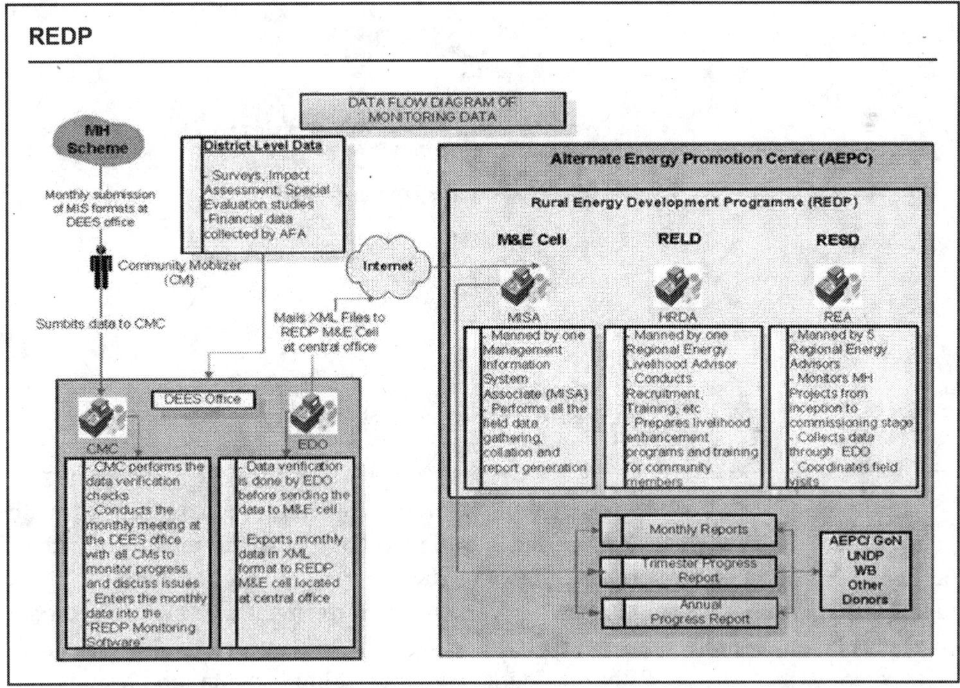

Under the existing process, the CM does a field level data collection of all the MH schemes under his purview and records the activities and progress in the prescribed formats at the end of each month. At present, there are 11 monitoring formats for data collection. The CM then submits the filled formats to the CMC in the DEES office. On completion of the data collection and verification process, a review meeting is organized by CMC consisting of EDO, Technical Officer, and CMs. The EDO reviews all field data and mails it across to the M&E cell of REDP. In the present monitoring arrangement, each of the Rural Energy Advisor (REA) for the five zones, functioning under RESD, monitor and track the progress of each of the MH projects from the inception till the commissioning of the project. The EDO collects and reports to the RESD all information from each of the MH project under his jurisdiction on a monthly basis and when required. The M&E cell of REDP at the central office then consolidates and generates various reports on a monthly, trimesterly, and annual basis to cater to the reporting requirements of senior management and donor agencies such as the World Bank and United Nations Development Program.

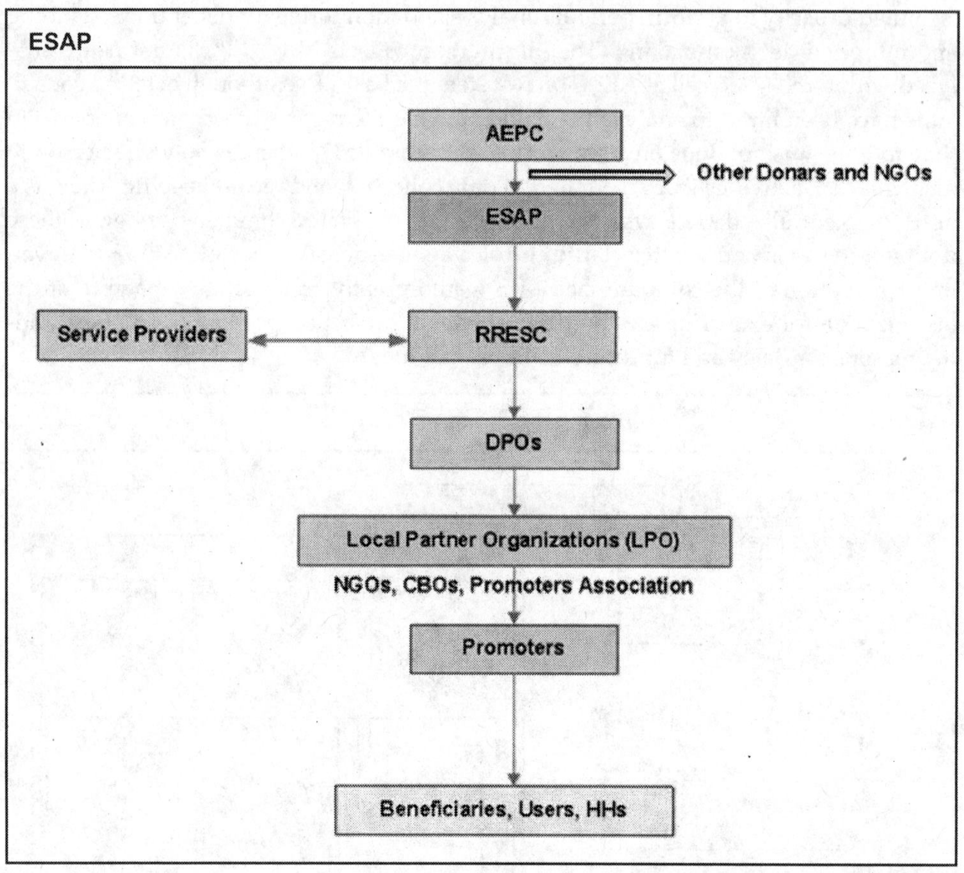

Under ESAP, the RRESC manages the field implementation in coordination with partner organizations at the district and local level. The Local Partner Organizations (LPOs) are responsible for implementation of the program at field level. The LPOs in turn select, train, support, and supervise promoters to implement the renewable energy component. The promoters are in direct contact with the community.

Presently, there are six RRESCs identified as project implementation partners and each RRESC cover around six to nine districts. The RRESCs are NGOs dedicated towards the empowerment and social inclusion of underprivileged groups. Each RRESC is also responsible for spreading awareness among rural poor on efficient renewable energy solutions and conduct different types of income-generating and skill development activities.

The implementation of RETs is determined based on an assessment of the demand and viability by LPOs. For MH installations implemented under the MGSP program of ESAP, the communities approach RRESC. With guidance and assistance of RRESC, a detailed feasibility study is conducted and sent to ESAP for approval and clearance. Once the project is approved, the user community applies to ESAP for subsidy and an agreement containing the project implementation schedule and disbursal mechanism is accordingly signed.

Monitoring of ESAP programs is done at different stages of the program and at different levels. Each installation is surveyed after a month of completion of the installation process. The testing is done jointly by the LPO and the promoter. After three to four months, a technical evaluation is conducted by the LPO and RRESC staff. Installations are selected for evaluation based on random sampling of 25 percent of the total installations. Quarterly reports are sent by RRESC to AEPC for approval. There is a provision for mid-term review of the annual work plan to assess the performance of the project against the program's objectives. The RRESCs perform an evaluation of the LPOs to check whether the targets are met and take feedback from them based on the interaction with the local communities. Apart from the evaluations, the field technical coordinator from the RRESC office goes to the field with the LPOs to collect information on the predefined formats. The reports are submitted to AEPC on a monthly basis.

Biogas

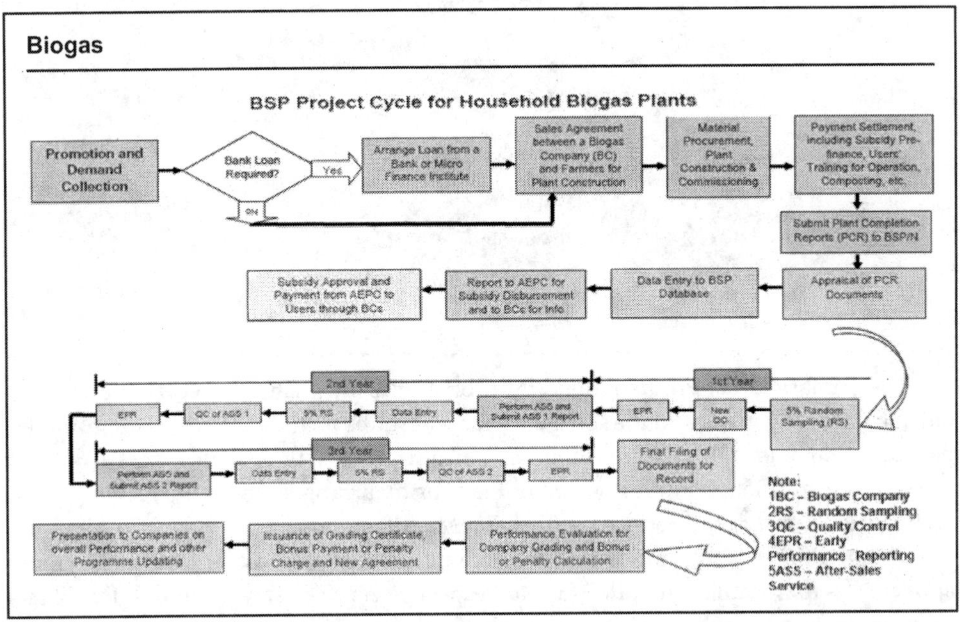

The implementation and monitoring process involves the public and private partners, including the NGOs and Biogas construction companies. The demand by the farmers, i.e., Biogas users, to set up a Biogas plant is created through awareness-building programs by the active involvement of NGOs, Cooperatives, Biogas Companies Network, MicroFinance Institutions, and Nepal Biogas Promotion Association (NBPA). A sales agreement is then executed between a qualified Biogas company and farmers for construction. The Biogas company provides the service for the construction, after-sales service, and user training. Biogas Support Program (BSP/N) coordinates with the Biogas companies through Nepal Biogas Promotion Center (NBPC), which is the umbrella organization of the Biogas companies. NBPC implements the promotion and training activities. BSP/N and recognized Biogas companies enter into an annual agreement every year. On completion of the construction activities, the Biogas companies submit the Plant Completion Reports (PCR) to BSP/N. The District Energy and Environment Unit (DEEU) carries out field verification of all the Biogas plants. BSP/N sends monthly progress reports to AEPC for subsidy disbursement.

AEPC approves the subsidy amount and payment is made by AEPC to the users through the Biogas companies. BSP/N also performs a quality control through a random sampling of 5 percent of the plants for performance reporting within the guarantee period. After one year of the construction, BSP/N performs a quality control through a random sampling of 2.5 percent of the plants for performance reporting. Similarly, DEEU monitors Biogas plants by random sampling of 2.5 percent of the plants annually, reporting to AEPC for verification, investigation, and action.

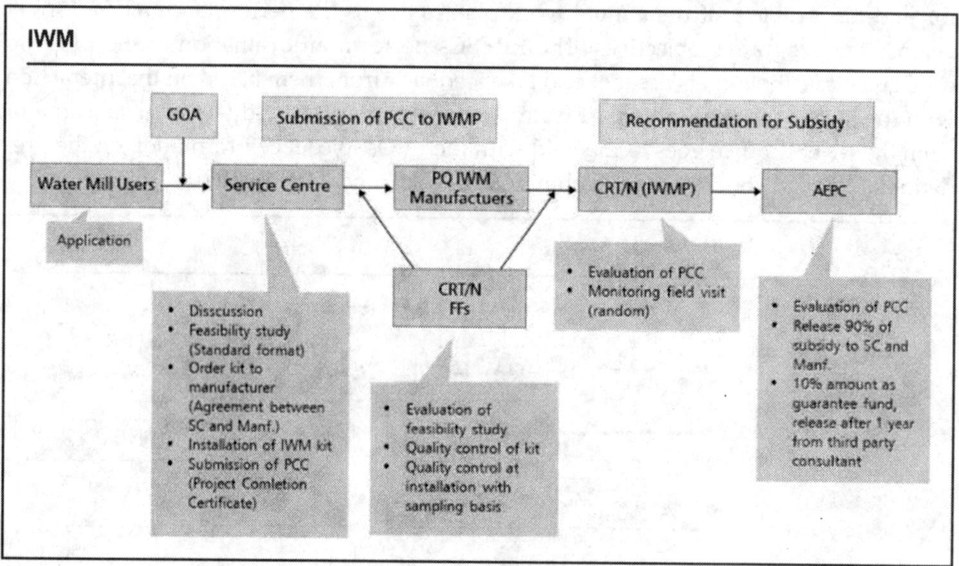

The primary target group of the IWM programs is the ghatta owners (millers) of the hill and mountain areas and the secondary target groups include the rural communities, particularly women who spend a lot of time in agroprocessing activities. In case of IWM projects, the installation and provision of the subsidy arrangement to the ghatta owners is channeled through the Service Centers and Manufacturers.

The traditional water mill users who are interested in improving their mills submit applications for installation of IMW to the respective service centers through the Ghatta Owners Association (GOA). GOA forwards the demand to a service center along with the required information. The service center studies the information, discusses the business plan with the applicant, and then conducts a feasibility survey. The service center sends the feasibility study report to the IWMP for approval. After assessment of the report, the IWMP approves the service center to install the IWM.

Under the approved scheme of installation, including the subsidy and credit matters, the service center then places order for the kit to the qualified manufacturer and also signs an agreement with the manufacturer. The Field Facilitators (FFs) of CRT/N review the feasibility study format and revert back to the service centers. CRT/N also performs regular quality control of the IWM installations on a random sampling basis. After receiving the order, the manufacturer supplies the IWM kit to the service center. The service center installs the IWM kit and other accessories at the site and submits the Project Completion Certificate (PCC) to CRT/N (IWMP). IWMP then evaluates the PCC and recommends to AEPC the release of the subsidy amount. Based on recommendation from IWMP, AEPC approves the subsidy and sends it to the mills' owners through their respective service centers and to manufacturers. After installation, the service center provides training on repair and maintenance and other aspects to the mill owners. The center also visits the installation sites to monitor performance of the system on a random basis.

Subsidy is delivered by AEPC through local banks to manufacturers and service centers. The subsidy provisions cover almost half of the total cost of the project. After the installation of the IWM, approximately 10 percent of the installations are surveyed to assess the socioeconomic and environmental benefit. The IWM program has in-house software to capture this information on a regular basis to quantify the impact of the program. The FF's CRT/N monitors 25 percent of the total installations at the gate of the manufacturers before the delivery to the service centers and during installation in the field on the basis of random sampling. Field visits are conducted frequently to monitor the effectiveness of the IWM units. The focus is on monitoring both the technical and social aspects.

ANNEX 2:

Roles and responsibilities of AEPC programs

Roles and Responsibilities for REDP Program

Designation	Existing Roles & Responsibilities of Staff	Proposed Roles & Responsibilities of Staff
Community Mobilizer	• Implements the community mobilization package in the VDC. • Collects the field-level data from the schemes. • Prepares weekly report and sends it to SO. • Mobilizes CO members to attend meetings, make savings, and undertake economic activities through credit. • Mobilizes household members to form COs. • Facilitates household-level socioeconomic development plans.	• Apart from data collection in the existing MIS formats, training shall be provided for data collection as per the requirement under the new formats. • Collect household level energy related baseline survey.
Community Mobilizer Coordinator	• Compiles all field data at DEES. • Reports to DCC office on how CMs work and status of VDCs. • Conducts SO review meetings—highlights issues/problems to DDC office. • Coordinates all CMs for implementing projects. • Gives feedback on how to conduct community mobilization and suggests any improvement measures.	• Apart from fulfilling the existing responsibility, coordinate the data collection and data entry and upload in the revised MIS formats. • Along with the EDO, analyze the program data to identify areas for further improvement.
Energy Development officer	• Reviews of all field data and information before data is sent to central office. • Monitors the projects under his/her jurisdiction to review whether they are functioning efficiently and identifies problem areas. • Gives feedback to the central office. • Manages all administrative, technical, and financial matters of the office. • Instructs CM, CMC, and TO. • Coordinates with line agencies, stakeholders, and donor agencies operating in the district.	• Checks data inconsistency issues and coordinates with concerned department to resolve the same. • Performance analysis of all ongoing projects. • Initiate proposals for new projects/work areas based upon analysis of data through MIRMS on problematic areas. • Suggest and recommend improvement areas for the ongoing projects.

Source: AEPC and Authors.

Roles and Responsibilities for REDP Program

Designation	Existing Roles & Responsibilities of Staff	Proposed Roles & Responsibilities of Staff
Management Information System Associate (MISA)	• Coordinates and facilitates the acquisition of the monitoring data from the DEES offices on a regular basis. • Manages the data upload process and ensures that complete data are available from all program districts. • Sends reminder to field agencies if the data are incomplete or late. • Prepares the monthly MIS reports for circulation to the top management. • Collects data, such as training status, project initiation, and completion status, from other teams such as Rural Energy Livelihoods Development (RELD) and RESD.	• Coordinate with other departments for data collection and updating of the MIRMS system. • Provide guidance and supervision to data entry operators for entering project and program data through DPR, Program documents, Project progress, etc. • System administration. • Liaison with third-party software and hardware vendors for all AMC related issues. • Administer the MIRMS system and configure and prepare reports in the system based on the reporting requirements of other line agencies within REDP.
Monitoring and Communication Officer (MCO)	• Mainly responsible for leading the team for supporting the National Program Manager (NPM) and other department heads. • Designing and implementing systematic documentation and dissemination of information for achieving the program outputs as appropriate. • Monitoring, evaluating, and reporting on the progress, achievements, and outcomes of the programs. • Performs analysis of program data on a regular basis to highlight areas where REDP should focus its attention for improving the program effectiveness. • Provide advisory support to partner organizations to strengthening the monitoring systems, data collection, compilation, and reporting based on analysis and recommendations.	• Align KPIs with strategic mission and vision of REDP and establish benchmarks. • Monitor the KPIs on a regular basis to measure current performance against goals and benchmarks to achieve continuous improvement. • Set targets in the system as outlined in program documents to measure performance. • Establish reporting needs. • Conduct interviews with stakeholders who will be impacted and find out which areas are effective or lacking and list the stakeholders for improvement and priorities. • Change, modify, or remove existing KPIs against whom desired objectives have been met and performance need not to be monitored; add new KPIs against any new objective that is decided.

Source: AEPC and Authors.

Roles and Responsibilities for ESAP Program

Designation	Existing Roles & Responsibilities of Staff	Additional Roles & Responsibilities of Staff
RRESCs	• Conduct feasibility assessment to find out the requirements of implementation of various RETs. • Survey of each installation after a month of the installation process. • Perform technical evaluation after three to four months of installation. • Conduct mid-term review of annual work plan to check whether program objectives are met.	• Apart from recording the activities and progress for each of the ESAP programs every month, the RRESC staff shall also capture data for the new MIS formats. • MIRMS to provide RRESC staff the targets set by the management for each of ESAP programs. • Perform regular analysis of achievement data to ensure that the program objectives are met.
Monitoring & Communication Officers (AEPC)	• Conduct monitoring of installation through independent consultants. • Random sampling of installations of various RETs for monitoring and verification. • Field verification of installations upon receipt of complaints. • Monitoring, evaluating, and reporting to program heads on the progress, achievements, and outcomes of the programs. • Monitoring of program and performance of the implementing agency.	• Apart from fulfilling the existing responsibility, should also ensure the regular uploads of the extracts of program data from respective ESAP databases to MIRMS database. • Coordinating and populating MIRMS database for MIS formats that is not captured in ESAP databases. • Utilize MIRMS to perform analysis and review of programs to ensure that the programs are meeting the intended objectives. • Compile results into meaningful reports and submit to top management for action.

Source: AEPC and Authors.

Roles and Responsibilities for Biogas Program

Designation	Existing Roles & Responsibilities of Staff	Proposed Roles & Responsibilities of Staff
Quality Controller Officer	• Facilitates the activities in the district as per the approved plan. • Link, coordinate, and network with NGOs and Biogas companies in the district. • Organize and/or facilitate different training and workshops in the district as per the plan. • Appraisal of PCR documents. • Control quality of accessories as per the agreed quality standards prescribed. • Send monthly report to the BSP.	• Provide advisory support to users and strengthen the monitoring systems and data collection, compilation, and reporting based on analysis and recommendations.
Monitoring and Communication Officer (MCO)	• No dedicated MCO appointed to coordinate the MIS consolidation and generation process.	• Appoint a dedicated MCO to performs analysis of program data on a regular basis to highlight areas where Biogas should focus its attention for improving the program effectiveness. • Align KPIs with strategic mission and vision of Biogas program. • Monitor the KPIs on a regular basis to measure current performance against goals and benchmarks to achieve continuous improvement. • Set targets in the system as outlined in program documents to measure performance. • Establish reporting needs. • Change, modify, or remove existing KPIs against whom desired objectives have been met and performance need not to be monitored; add new KPIs against any new objective that is decided.
Senior Associates	• Perform the quality monitoring and coordinating functions for BSP. • Conduct quality checks during and after the construction of the plant based on random sampling of Biogas plants.	• Performance analysis of all ongoing projects. • Initiate proposals for new projects/work areas based upon analysis of data through MIRMS on problematic areas. • Suggest and recommend improvement areas in the functioning of the ongoing projects.

Source: AEPC and Authors.

Roles and Responsibilities for IWM Program

Designation	Existing Roles & Responsibilities of Staff	Proposed Roles & Responsibilities of Staff
Ghatta Owners Association (GOA)	• Performs all the awareness campaigns. • Provides their services to the mill owners.	• Management training to be provided to the GOAs. • Focused training to be provided on credit mobilization, technology development, and marketing of IWM-related products to assist the mill owners and the end users. • Develop as a rural energy service center.
Service Centers	• Social mobilization for the marketing and promotion of IWM. • Demonstration and awareness visits for creating awareness. • Facilitate the formation of mill owners' group to increase the capacity for institutionalization of the millers. • Conduct feasibility study for installation of IWM. • Procurement of IWM kits from prequalified manufacturers. • Linking available program subsidies to IWM owners. • Installation of IWM. • Provision of repair and maintenance services.	• Undertake periodic repairs and maintenance programs to provide efficient services to mill owners so that the IWM runs in a sustainable manner with less breakdown time.
Manufacturers	• Production of quality kits as per standards set by the program. • Facilitate CRT/N for quality checking of the products.	
Account Officer	• Release fund to district units as per the advance request in line with the approved annual plan. • Prepare monthly financial report to submit to Netherlands Development Organization/Nepal, including bank reconciliation statements. • Prepare financial reports to submit to AEPC as per the available formats and upon request. • Assist subsidy channeling to SPs for AEPC. • Provide assistance to program director and manager.	• The officer is expected to perform the same set of activities.
Gender Expert	• Transfer of gender concept to Ghatta owners, service centers, and other stakeholders. • Conduct gender assessment in IWM program areas. • Perform gender analysis, which explicitly identifies differences between women and men regarding access to and control over income and participation in decision-making and its benefits and the direct and indirect impact on the household. • Preparation of gender guideline and indicators for IWM program.	• Perform gender impact analysis of all IWM projects within the jurisdiction. • Initiate proposals for new projects/work areas based upon analysis of data through MIRMS on problematic areas. • Suggest and recommend improvement areas in the functioning of the ongoing projects.

Source: AEPC and Authors.

Roles and Responsibilities for IWM Program

Designation	Existing Roles & Responsibilities of Staff	Proposed Roles & Responsibilities of Staff
Field Facilitators (FFs)	• Facilitates the activities in the district as per the approved plan. • Link, coordinate, and network with NGOs, INGOs, and ghatta owners in the district. • Support service centers during the process of installation of IWM. • Organize and/or facilitate different training and workshops in the district as per the plan. • Monitor the feasibility study and IWM installation and support finalization of PCC. • Control quality of kit at manufacturer's door as per the agreed quality standards for IWM kits. • Send monthly report to the center. • Promote rural/renewable energy technology around IWM area.	• Provide advisory support to mill owners, GOAs, and service centers to strengthen the monitoring systems and data collection, compilation, and reporting based on analysis and recommendations.
Monitoring and Communication Officer (MCO)	• Mainly responsible for coordinating the MIS data collection and consolidation to meet the reporting requirements of program manager and top management. • Monitoring, evaluating, and reporting on the progress, achievements, and outcomes of the programs. • Performs analysis of program data on a regular basis to highlight areas where IWM should focus its attention for improving the program effectiveness.	• Align KPIs with strategic mission and vision of IWM program. • Monitor the KPIs on a regular basis to measure current performance against goals and benchmarks to achieve continuous improvement. • Set targets in the system as outlined in program documents to measure performance. • Establish reporting needs. • Change, modify, or remove existing KPIs against which desired objectives have been met and performance need not to be monitored; add new KPIs against any new objective that is decided.

Source: AEPC and Authors.

ANNEX 3:

Sample description for household and enterprise survey

The purpose of the survey is to assess the development impacts of the micro-hydro schemes in rural Nepal that were implemented under REDP. The rural household and enterprise survey was conducted jointly by the World Bank and AEPC in Nepal during early 2009. It also uses a wide range of district-level information that is publicly available to complement the survey data. Furthermore, it draws on reports and presentations prepared by AEPC and the World Bank. This Annex briefly describes the survey instruments.

Nepal has three physiographic regions – *Terai* (plains), *Hill* (hills), and *Parbat* (mountains) – parallel to one other, extending roughly from the northwest to southeast borders of the country. Terai region is the lowland tropical area in the south with an elevation of 1,000 to 3,000 feet. Next to Terai is the Hill region (called *Pahar* in Nepali) with an elevation of 4,000 to 13,000 feet. Finally, the Mountain region (called Parbat in Nepali) is characterized by inclement weather and the roughest terrain with an elevation of 13,000 feet to 29,000 feet. Among these regions, the Hill and Mountain regions are not suitable for grid extension because of the altitude and terrain condition, and MH installations have almost exclusively been implemented there.[1] So, these two regions were considered for the survey of this study. However, during the pretesting of the questionnaire, households in the Mountain region were found extremely inaccessible and therefore eventually excluded from the survey.[2] As such, survey was confined to the Hills region.

Nepal has also five development regions and the survey covered all of them: East, Central, West, Mid-west, and Far-west. Among the twenty-five districts in the Hill region, twelve were selected randomly for this study, and one VDC (Village Development Committee) was selected from each district.[3] Finally, twenty-four MH installation areas were selected, from which households were selected. In the selection of MH locations, the following considerations were made.

- A minimum of three MH installation locations were selected from each of the development regions, and more from regions where a large number of MH plants are established.
- Only those MH locations were selected where MH schemes had been established for more than three years, allowing enough time for benefit accumulation.
- Locations with very low generation capacity were excluded. More specifically, only those MH locations with a generation capacity of at least 7 kW were selected.

With above considerations, sample households (both for treated and control groups) were selected based on the probability proportionate to size (PPS) sampling method. Since the focus of this study is MH, MH households were purposely oversampled. Control households were selected from both MH locations and non-MH locations. All in all, 2,500 households were selected, of which 1,500 are treated, 571 are control from MH locations, and 429 are control from non-MH (control) locations. Table A1 shows the sample distribution.

Table A1: Household distribution in the sample

District	Development region	Treated households	Control households in		Total households
			Treatment locations	Control locations	
Tehrathum	East	172	65	0	237
Okhaldhunga	East	60	23	83	166
Kavre	Central	265	102	60	427
Tanahun	West	78	30	0	108
Myagdi	West	122	47	0	169
Parbat	West	33	13	0	46
Baglung	West	357	133	88	578
Pyuthan	Mid-west	91	35	0	126
Dailekh	Mid-west	76	29	66	171
Achham	Far-west	75	29	0	104
Dadeldhura	Far-west	22	8	0	30
Baitadi	Far-west	148	56	131	335
12 districts	5 regions	1,499	571	429	2,497

Source: World Bank-AEPC survey 2009.

Besides households, forty-seven microenterprises were selected from the same locations as the households were selected from and their selection followed similar criteria. Table A2 shows the distribution of the sample microenterprises by district.

Table A2: Enterprise distribution in the sample

District	Development region	Treated enterprises	Control enterprises in		Total enterprises
			Treatment locations	Control locations	
Tehrathum	East	3	3	0	6
Okhaldhunga	East	3	0	5	8
Kavre	Central	6	3	4	13
Tanahun	West	4	0	0	4
Myagdi	West	3	0	0	3
Parbat	West	1	2	0	3
Baglung	West	6	4	5	15
Pyuthan	Mid-west	2	0	0	2
Dailekh	Mid-west	1	1	4	6
Achham	Far-west	1	0	0	1
Dadeldhura	Far-west	1	0	0	1
Baitadi	Far-west	3	0	5	8
12 districts	5 regions	34	13	23	70

Source: World Bank-AEPC survey 2009.

Notes

1. Out of the forty districts that have MH plants under REDP only two are in Terai region.
2. Because of the high altitude, inclement weather, and rugged topography, only 8 percent of the country's population live in Mountain region.
3. A VDC is an administrative unit below the district.

ANNEX 4:

Propensity score matching technique

Propensity score matching (PSM) is a useful technique to estimate program participation impacts when it is assumed that observed characteristics determine program participation (for example, household's access to MH). At the heart of this technique is *propensity score*, defined by the conditional probability of receiving a treatment given pretreatment characteristics: $p(X) \equiv \Pr\{D = 1 \mid X\} = E\{D \mid X\}$, where $D = \{0,1\}$ is the indicator of exposure to treatment and X is the multidimensional vector of pretreatment characteristics. Participants are matched with nonparticipants based on the propensity score. Average impact of the program on the participants (called *average treatment of the treated* or ATT) is then calculated as the difference in mean outcomes between the two groups, or formally: $T = E\{E(Y_{1i} \mid D_i = 1, p(X_i)) - E(Y_{0i} \mid D_i = 0, p(X_i)) \mid D_i = 1\}$,
where γ_{li} and γ_{0i} are the potential outcomes of treatment and no treatment, respectively. For more on PSM technique, see Rosenbaum and Rubin (1983); Ravallion (2008); and Khandker, Koolwal, and Samad (2009).

The basic steps in PSM estimation areas are presented below:

1. Estimating a model for program participation

First, participation in the treatment (*T*) is estimated on all the observed control variables (X) that are assumed to determine participation. Since *T* = 1 for those who participate in the treatment and 0 otherwise, either a probit or logit model is used for program participation. After the participation equation is estimated, the predicted value of *T* is calculated. This predicted outcome is the estimated probability of participation in the treatment, which is also called the propensity score. Every observation in the sample will have an estimated propensity score, represented formally by: $\hat{P}(X \mid T = 1) = \hat{P}(X)$.

2. Defining the region of common support and balancing tests

Region of common support is where distributions of the propensity score for treatment and control groups overlap. Nonparticipants that are not with the common support region are discarded. Then the treatment and comparison groups are balanced in such a way that similar propensity scores are based on similar observed X. Balancing is needed because a treated group and its matched nontreated comparator might have the same propensity scores but not necessarily the same observationally. Formally, one needs to check if $\hat{P}(X \mid T = 1) = \hat{P}(X \mid T = 0)$.

Balancing tests can also be conducted to check whether, within each quintile of the propensity score distribution, the average propensity score and mean of X are the same.

3. Matching participants with nonparticipants

After balancing is done, a suitable matching technique is adopted to pair participants with nonparticipants. It requires calculating a weight for each matched participant–nonparticipant set. As discussed below, the choice of a particular matching technique may therefore affect the resulting program estimate through the weights assigned.

Nearest-neighbor matching: Each treatment unit is matched to the comparison unit with the closest propensity score. One can also choose *n* nearest neighbors and do matching (usually $n = 5$). Matching can be done with or without replacement. Matching with replacement, for example, means that the same nonparticipant can be used as a match for different participants.

Caliper or radius matching: A threshold or "tolerance" is imposed on the maximum propensity score distance. This procedure involves matching with replacement only among propensity scores within a certain range. A higher number of dropped nonparticipants may happen, thereby increasing the chance of sampling bias.

Stratification or interval matching: This procedure partitions the common support into different intervals and calculates the program's impact within each interval. Within each interval, the program effect is calculated from the mean difference in outcomes between treated and control observations. The overall program impact is the weighted average of these interval impact estimates, where weight is the share of participants in each interval.

Kernel and local linear matching: Nonparametric matching estimators such as kernel matching use a weighted average of all nonparticipants to construct the counterfactual match for each participant. If P_i is the propensity score for participant *i* and P_j is the propensity score for nonparticipant *j*, the weights for kernel matching are given by:

$$\omega(i,j)_{KM} = K\left(\frac{P_j - P_i}{a_n}\right) \Big/ \sum_{k \in C} K\left(\frac{P_k - P_i}{a_n}\right),$$

where $K(\cdot)$ is a kernel function and a_n is a bandwidth parameter.

4. Calculating average treatment of the treated (ATT)

The average treatment (ATT) of program intervention on an outcome *Y* can be obtained as:

$$ATT = \sum_{i=1}^{N_T} (Y_{ij}^T - \sum_{j=1}^{N_C} W_{ij} Y_{ij}^C) / N_T ,$$

where Y_{ij}^T and Y_{ij}^C are the outcomes for program participants and nonparticipants, respectively; N_T is the number of participants; N_c is the number of matched nonparticipants; and W_{ij} is the associated propensity score-based weight given to the control observations in matching with the participants.

As mentioned in the sample description, MH households have been overdrawn in the sample. To make the findings of this study representative of rural Nepal, all analyses have been weighted by the actual distribution MH users and nonusers at the district level.

ECO-AUDIT

Environmental Benefits Statement

The World Bank is committed to preserving endangered forests and natural resources. The Office of the Publisher has chosen to print World Bank Studies and Working Papers on recycled paper with 30 percent postconsumer fiber in accordance with the recommended standards for paper usage set by the Green Press Initiative, a non-profit program supporting publishers in using fiber that is not sourced from endangered forests. For more information, visit www.greenpressinitiative.org.

In 2011, the printing of these books on recycled paper saved the following:

- 6 trees*
- 2 million Btu of total energy
- 596 lb. of net greenhouse gases
- 2,690 gal. of waste water
- 170 lb. of solid waste

* 40 feet in height and 6–8 inches in diameter